LOW PROFILE AMATEUR RADIO

Operating a **Ham Station** from Almost Anywhere

Second Edition

Al Brogdon, W1AB

Published by:
ARRL *The national association for* AMATEUR RADIO

Production ———————

Michelle Bloom, WB1ENT

Sue Fagan — Front Cover

Jodi Morin, KA1JPA

Joe Shea

David Pingree, N1NAS

Table of Contents

Foreword

Here we are in the early years of the 21st century. Many owners of single-family homes or condominiums and renters of single-family homes or apartments must comply with the requirements placed on them by their deeds or rental agreements — the dreaded CC&R (Covenants, Conditions and Restrictions). Virtually all CC&R requirements prohibit outside antennas.

How is a ham to get on the air, especially one who wants to operate on the HF bands? Don't despair — it can be done. This book will present information to help you learn how to install HF antennas within your CC&R limitations — antennas that will work quite well. This book will discuss, for the most part, HF antennas and operating, and will tutor you on experimentation using cut-and-try techniques.

In addition to looking at low-profile antennas for HF, this book details other aspects of low-profile hamming, such as modes of operation, transmitter output power, interference possibilities, VHF/UHF operation, etc.

I hope this book will help you learn how you can enjoy Amateur Radio, even under severe restrictions. It's not as difficult as you might first think...

David Sumner, K1ZZ
ARRL Chief Executive Officer

About the ARRL

The seed for Amateur Radio was planted in the 1890s, when Guglielmo Marconi began his experiments in wireless telegraphy. Soon he was joined by dozens, then hundreds, of others who were enthusiastic about sending and receiving messages through the air—some with a commercial interest, but others solely out of a love for this new communications medium. The United States government began licensing Amateur Radio operators in 1912.

By 1914, there were thousands of Amateur Radio operators—hams—in the United States. Hiram Percy Maxim, a leading Hartford, Connecticut inventor and industrialist, saw the need for an organization to band together this fledgling group of radio experimenters. In May 1914 he founded the American Radio Relay League (ARRL) to meet that need.

Today ARRL, with approximately 150,000 members, is the largest organization of radio amateurs in the United States. The ARRL is a not-for-profit organization that:

◆ promotes interest in Amateur Radio communications and experimentation

◆ represents US radio amateurs in legislative matters, and

◆ maintains fraternalism and a high standard of conduct among Amateur Radio operators.

At ARRL headquarters in the Hartford suburb of Newington, the staff helps serve the needs of members. ARRL is also International Secretariat for the International Amateur Radio Union, which is made up of similar societies in 150 countries around the world.

ARRL publishes the monthly journal *QST*, as well as newsletters and many publications covering all aspects of Amateur Radio. Its headquarters station, W1AW, transmits bulletins of interest to radio amateurs and Morse code practice sessions. The ARRL also coordinates an extensive field organization, which includes volunteers who provide technical information and other support services for radio amateurs as well as communications for public-service activities. In addition, ARRL represents US amateurs with the Federal Communications Commission and other government agencies in the US and abroad.

Membership in ARRL means much more than receiving *QST* each month. In addition to the services already described, ARRL offers membership services on a personal level, such as the ARRL Volunteer Examiner Coordinator Program and a QSL bureau.

Full ARRL membership (available only to licensed radio amateurs) gives you a voice in how the affairs of the organization are governed. ARRL policy is set by a Board of Directors (one from each of 15 Divisions). Each year, one-third of the ARRL Board of Directors stands for election by the full members they represent. The day-to-day operation of ARRL HQ is managed by a Chief Executive Officer.

No matter what aspect of Amateur Radio attracts you, ARRL membership is relevant and important. There would be no Amateur Radio as we know it today were it not for the ARRL. We would be happy to welcome you as a member! (An Amateur Radio license is not required for Associate Membership.) For more information about ARRL and answers to any questions you may have about Amateur Radio, write or call:

ARRL—The national association for Amateur Radio
225 Main Street
Newington CT 06111-1494
Voice: 860-594-0200
Fax: 860-594-0259
E-mail: **hq@arrl.org**
Internet: **www.arrl.org/**

Prospective new amateurs call (toll-free):
800-32-NEW HAM (800-326-3942)
You can also contact us via e-mail at **newham@arrl.org**

1

Introduction

Let me introduce myself, so you can get a sense of where my ideas, opinions, and suggestions are coming from. I was first licensed as WN4UWA (one of the early Novices) on January 15, 1952. Three months later, I got my General class license and, in May 1956, the Amateur Extra class. From early on, I read *QST* and the ARRL publications to learn about Amateur Radio, in addition to hanging out with experienced hams to learn from them and to see what antennas they used and how they operated their stations.

Beginning in those early years and continuing all through my ham career, I experimented with antennas, sometimes following ideas from *The ARRL Antenna Book* and sometimes just trying things to see if they would work — sometimes they did, and sometimes they didn't. Each time I experimented with antennas, I learned a little more about how antennas worked. I learned which antennas worked well, which antennas didn't work, and why the good ones worked.

I reached this conclusion: Hands-on antenna experimentation is a good way to learn about the beasties, backed up with reading the antenna literature to understand why they do or do not work. I recommend that approach to you.

Why? Because the ham who is having to put up low-profile antennas usually cannot build an exact copy of an antenna that is described in the antenna literature or in the pages of *QST* or other magazines or books. Most of the antennas described in those sources are for the ham who can flagrantly throw a multitude of wires and feed lines into the sky, for all to see and for other hams to admire. Low-profile antennas must be hidden

Some Early Low-Profile Antenna Fun

During the mid-1950s I was studying electrical engineering at Tennessee Tech, in Cookeville, Tennessee, and signing W4UWA on the ham bands. David Knight, W4ZJY, was in high school at the time and lived a few miles from me. Dave and I became friends through our participation in the CW nets of the National Traffic System.

Dave and I often played a game on the air, each trying to outdo the other with silly antennas. We would make contact on 3635 kHz (the Tennessee CW Net frequency, then and now) using our normal antennas, and we would start trying all sorts of crazy antennas, one asking the other, "RST?", then the other taking a turn at transmitting, using a weird antenna.

Some of the antennas I remember using during those tests included the window screen on the window of my shack; the old-fashioned, open bedsprings on my bed; and a 20-inch whip. Neither of us ever tried an antenna that the other couldn't hear.

In addition to these games with Dave, I would sometimes make contact with other hams at long distances using those scruffy antennas. I wouldn't get very good signal reports, but I could make contacts over pretty good distances using antennas that were pretty bad.

Thus, 50 years ago I came to the realization that you could make contacts with minimalist antennas. And that's why I have never hesitated to use less than optimal antennas. They work — perhaps not as well as a high dipole, but they work well enough to get you on the air! Many times I've used pretty bad antennas to operate on 160 meters, yet I have made plenty of contacts. In that case, I give full credit to the 160-meter CW operators as having some of the best ears in Amateur Radio — thanks, gentlemen!

Dave, W4ZJY, continued to give heavy support to NTS nets for the ensuing decades, doing yeoman service with nets at all levels (Section, Regional, and Area) as well as net liaison and trunk-line work. I continued to support Section-level nets for many years. NTS provided good training for us young hams of that era. We learned to listen carefully, to copy well, and to follow instructions exactly — all of which are good habits for any ham.

or disguised. But you can take an antenna idea from the literature and adapt it to fit your space and visibility restrictions…if you know how to go about making those adaptations. The better you understand how antennas work, the better you can make those adaptations.

I've spent most of my ham operating time over the past 54 years on HF CW, but I've also operated HF SSB and VHF FM. For a few years in the latter 1960s I had a tower, a good two-band beam (two loaded elements on 40 and three full-size elements on 20), and a legal-limit amplifier. Except for that short foray into the ranks of the Medium Guns, I've operated with modest power (100 W or less) and simple wire antennas. I enjoy meeting other hams on the air for ragchewing, rather than today's more common "599 73" contacts. My opinion, which runs counter to the opinions of many of today's hams, is that you can greatly enjoy Amateur Radio by relaxing and making the most of what you can do, rather than straining yourself to reach for goals that might be beyond your grasp. That is especially true for the CC&R-inhibited ham.

Now that we have been properly introduced, let's get on with looking at low-profile antennas and operation.

2

Overview of This Book

Most of the problems hams have in setting up low-profile stations lie in the area of antennas for HF operation. Therefore, that will be the major emphasis in this book.

There will be some discussion of VHF and UHF antennas, but most VHF/UHF operators can install pretty good antennas in their attics for those higher bands. With low-profile HF antennas, your signals will likely be not as strong as the average signal. Therefore, this book will discuss operating practices and operating tips. Better operating practices can often make the difference between making a contact and missing it.

In this book I will often mention specific antennas and antenna tuners, as well as other pieces of equipment. The particular manufacturers and models mentioned are for illustrative purposes, and the reader should not take a citation to be a recommendation for that particular piece of equipment. Rather, when you decide on a type of equipment to buy (or build), do research in the amateur literature to find out what manufacturers' equipment (or published circuit) might work well for you. Read the "Product Reviews" in *QST*. Ask other hams you know for their opinions. Toss that all together and you will reach an informed decision about what particular model and manufacturer will serve you best.

3

CC&R—A Modern Dilemma

Because you're reading this book, it's likely that you already know what "CC&R" means. But, just to be sure: CC&R is the abbreviation for Covenants, Conditions and Restrictions. Those are the conditions and limitations that you agree to honor when you buy your home or when you rent or lease a condo or apartment.

When I received my first Amateur Radio license in 1952, most communities had no restrictions on ham antennas. Only a very few towns had antenna restrictions of any kind.

Over a half century has passed since then. More and more towns and housing developments have drawn up agreements to rigidly control the requirements, restrictions and use of a property. Robert H. Nelson took a current look at CC&R in his article, "Home Is Where the Rules Are," published in the *Washington Post* on Sunday, December 18, 2005. Here is a very pertinent observation from Nelson's article:

> *In 1970, only 1 percent of us lived in association-governed communities. Between 1980 and 2000, fully half of the new housing built in the United States was in such communities. In fast-growing parts of the country, especially in the South and the West, essentially all new development involves private community associations.*

And that's one of the reasons why today's hams are having problems and having to go into the low-profile mode. Nelson also reports that there were, at the time his article was written,

An Example of Enforcement

This story was told on the pages of the *Washington Post* in 1985. I followed the story with morbid interest as it developed over a period of several months. Here's a quick overview of the story.

Max Parsons, a Star Trek fan, lived in the CC&R'd community of Franklin Farm in Fairfax, Virginia, near Washington, DC. Parsons very much wanted to be able to receive satellite TV at his home, but his CC&R would not allow any type of outside antenna.

Parsons installed a small dish antenna for satellite TV reception that was hidden under a sawbuck picnic table's umbrella. Picnic tables and attached umbrella were allowed by the CC&R! And so the antenna/picnic table was placed in Parsons' backyard, with power and signal cables between the antenna and the house buried underground.

I remember that the photos in the *Washington Post* showed the umbrella to be slightly unusual in shape; its top was more rounded than a real picnic-table umbrella. The shape was necessary to hide the dish and the satellite TV receiver. Also, the sawbuck table wasn't of the usual proportions of a picnic table. But both were visually near enough to the real things to make Parsons think that he could get away with the subterfuge.

And he did, enjoying his satellite TV reception for a short time. But then the homeownder association realized that this was no ordinary picnic table....

A neighbor investigated and reported his findings to the association. The homeowners' association wrote Parsons and told him to remove the faux picnic table from his yard. Parsons declined, saying that it was a picnic table with umbrella, which was allowed under the CC&R.

The homeowners' association filed a lawsuit to force the removal of the picnic table. But, just before the suit went to court, the association amended its CC&R to permit satellite dishes if they are properly screened from view. Parsons agreed to screen his picnic table from view and the lawsuit was dropped.

Parsons then built an $8,000 second-story deck on the back of his split-level house and sank the electronic picnic table into a well in the deck, so no one could see it from the ground. That satisfied the homeowners' association, whose representative, incidentally, had said that "the picnic table was ugly." As always, beauty is in the eye of the beholder.

The reason I am relating this story to you is to illustrate that it is possible to come to compromises with homeowners' associations. But it isn't easy; it's expensive to do legal battle, and — as in this case — even coming to a compromise solution is expensive.

The only report I could find on the Web gives only the highlights of the story, and not the details that I read with fascination from month to month in the *Washington Post*, as the story unfolded. Go to **www.brincefield. com/dish2.htm** to read that abbreviated version of the story.

W3LPL and CC&Rs

Frank Donovan, W3LPL, has for many years maintained and operated a superb and highly competitive multi-multi contest station in Glenwood, Maryland. As you drive between Baltimore and Frederick on I-70, if you know when and where to look, you can see his four 200-foot towers, two 100-foot towers and 140-foot rotating tower.

Frank moved to Glenwood from his previous home in Annapolis, Maryland, one summer. One very interesting aspect of the move was that after he and his contest crew finished the winter/spring contesting season in Annapolis that year, the crew moved the ham station to the new home, erected the tower and antennas, assembled the station and were ready for the fall season's major contests. It was an almost superhuman effort that was conducted during a very hot summer, with the temperature and humidity both near 100 on many days. Those guys just about melted from the heat, but they finished the job on time! Bravo!

After Frank had moved to his new home and multi-multi contest station, so the story goes, a prospective homebuyer was looking at a house for sale that was within sight of Frank's enormous antenna field. The prospective buyer mentioned to the realtor that he wanted to put up a modest TV antenna tower so he could get good signals from both the Washington and Baltimore TV stations (this was before cable TV systems had spun their web into Frank's rural neighborhood).

The realtor replied that homeowners there were not allowed to put up antenna towers or outside antennas. Flabbergasted and quite annoyed, the prospective buyer asked, "Why not?"

"Because of the CC&R that apply to this neighborhood."

Pointing to the *very* obvious W3LPL antennas and towers, the prospective buyer said, "But look at the antennas that guy has!" The realtor then explained that Frank's home was not in the CC&R-controlled neighborhood where they were standing, and Frank didn't have a CC&R where he lived.

The *former* prospective buyer got in his car, slammed his door, drove off in a huff, and the realtor never saw him again. You know what realtors always say about property — "Location, location, Location!"

1,250,000 people on the boards of America's 275,000 homeowners' associations. One might suppose that, with that many homeowners' associations in existence, a few might be friendly toward hams and their antennas, but I have yet to hear of one that is.

In other words, the situation for hams is getting progressively worse. When you buy a property that has CC&R, *you sign away your rights to have, among other things, outside antennas.*

CC&R allow the homeowners living within the covered development, community or whatever to have *only* the things on their property that fit the community's idea of what is good and beautiful. Hams who live in communities or housing developments with CC&R have no choice but to comply. They must install and use only antennas (1) that are indoors and completely out of sight, (2) that are disguised as objects that *are* allowed (such as flagpoles), (3) that are virtually invisible, or (4) that can be put outdoors under cover of darkness, and brought back inside before the dawn's early light. Your home might be your castle, but its ramparts and towers must not conflict with the restrictions that have been placed thereon.

How bad can CC&R be? I know of a town in Maryland that has an approved list of the plants, flowers and bushes that may be planted in the homeowners' yards. Another CC&R list in that town tells what shades of what colors can be used to paint the outside of your home. Still another list tells what objects you can have installed or sitting in your yard. I know this because one of my sons lived in that town.

Anything *not* on those lists is automatically excluded. Of course, as you would guess, none of these lists have "Amateur Radio antenna" on them. What is the poor ham to do if he or she lives in a community that has severe CC&R restrictions? That's easy — he or she (1) reads this book; (2) learns some basic things about antenna design, installation and use; (3) installs some good hidden or disguised antennas; and (4) enjoys operating on the bands and modes of her or his choice.

Now that I've finished venting about CC&Rs, let's get on with learning about antennas and how to make them play well, while being invisible...or nearly so.

4

My Favorite Low-Profile HF Antenna Systems

When operating on HF in a low-profile situation, the antenna is usually the most difficult part of the equation. Therefore, let's now take a quick overview of what I consider my favorite low-profile HF antennas.

I could have put these recommendations at the end of the book, in a clever attempt to get the reader to read the entire book before getting to the conclusions. But then I realized that most readers would skip the middle of the book and go direct to the recommendations. So here they are, right up front. Further discussion of these various antenna systems (and how to best use them) will be presented in later chapters.

My recommendations are made mainly on ease of operation and reliability. For the better choices, you will have to spend a bit more money than for the lower-ranked choices… but isn't that true about everything?

Here are my choices, in order of preference.

MY FIRST CHOICE

My first choice for a low-profile HF antenna is a wire antenna system fed via an automatically tuned antenna tuner such as the SG-230 Smartuner, by SGC. The SG-230 covers 1.8 to 30 MHz and is rated at 200 W PEP or 80 W continuous duty. You can use it for the typical 100-W CW rig if you remember to not keep the key closed at full output for more than a few seconds at a time. For the tuner's full specs, go to **www.sgcworld.com/230ProductPage.html**.

The manufacturer's suggested retail price (MSRP) is $595; you can likely get it at a slightly reduced price from your favorite ham store.

If you want to have a larger safety margin by using a higher wattage rating, you can buy the SG-235, which is similar to the SG-230 but rated at 500 W PEP and 200 W continuous. For that tuner's technical specs, go to **www.sgcworld.com/235ProductPage.html**. The extra money you pay for the increased power rating is about a kilobuck — the SG-235's MSRP is $1595. But remember that paying a higher price will

Figure 1 — The SGC SG-237, a typical commercial automatic antenna tuner. Designed for outdoor use, it is completely waterproof and can be mounted at the end of the coaxial feed line.

give you a larger safety margin with your 100 W transceiver when operating CW.

Another SGC tuner you might consider is the SG-239 tuner, budget priced at $180. It is also rated at 200 W PEP and 80 W continuous. Read about it at **www.sgcworld.com/239ProductPage.html**.

In Chapter 7 we will discuss the wire-antenna arrangements that will play well with the SGC tuners. Note that the SGC autotuners are typically used in a location that is remote from the transceiver. Most of the SGC autotuners are enclosed in weatherproof cases, so they can be outside without any environmental problems. The autotuner is connected to the transceiver by a run of coaxial cable, and the wire antenna connects directly to the tuner's terminals.

All of the SGC autotuners require 12 V dc for their operation. There are various ways to get 12 V dc to the tuner, but the slickest way is to run the dc voltage through the coaxial cable between the rig's output and the tuner. All you need to do is to have a way to isolate the DC from the RF at each end of the coax, an old trick. A good practical description of how the ham can do it was presented by Phil Salas, AD5X, in "Remote DC Power through Your Coax," in the July 2004 issue of *QST*.

After you set up the antenna, the autotuner, and the run of coax between the two, all you need do to get a resonant antenna system is to feed a small RF signal down the coax to the autotuner, and the autotuner quickly works its magic, bringing the antenna system into resonance within 10 ms or so. When you change bands, the autotuner once again retunes quickly, and you are ready to operate on the new band. Furthermore, the autotuner remembers its settings for each band and goes to them automatically when you transmit a small RF signal down the line. Thus, returning to a previously used frequency will be even faster than the first time you used that frequency.

A caution: Don't think you can use the autotuner that's built into your rig (or any autotuner that will tune over a limited range of impedances) for the typical low-profile antenna system. Those autotuners are designed to match antenna-system impedances over a typical range of 30 to 120 Ω. A typical low-profile antenna system might present an impedance value as high as 2000 Ω on some bands. Therefore, the limited-range autotuners won't cut it. Read the specs carefully before buying an autotuner to use with a compromise antenna, such as many of the low-profile antennas. Make sure it tunes over a wide range of impedances.

Once you have set up an autotuner with a given antenna, you usually don't have to do anything more about the antenna system. That's why this arrangement is my first choice for low profile operating.

MY SECOND CHOICE

My second choice of a low-profile HF antenna is a simple variation on the first choice. It also uses an autotuner, but this time the tuner is placed near the transceiver, with a short length of coax connecting the two, and ladder line between the autotuner and the antenna.

Both SGC and MFJ autotuners are good choices when the autotuner is near the rig rather than in a

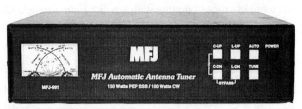

Figure 2 — The MFJ-991 is an example of an automatic antenna tuner designed for indoor use.

remote location. The MFJ autotuners have meters and controls on their panels, so they must be within arm's reach of the operator. Some MJF autotuners you can consider for your low-profile antenna system include the MFJ-991B (or the earlier MFJ–991 model), MFJ-993B (or MFJ–993), MFJ-994B (or MFJ-994), and the MFJ-929.

As with the SGC auto-tuners, you must size the MFJ autotuner to fit the power output you expect to use. One interesting note: MFJ calls their MFJ-991B "the world's first dual-power-level tuner." You can run 300 W on SSB or CW and match 6 to 1600 Ω antennas, or run 150 W on SSB or CW and match the wider range of 6 to 3200 Ω.

In comparing the MJF autotuners with the SGC, you will see the obvious difference in pricing, especially with the higher-power models. I leave it to your judgment (and budget) to decide which way to go.

In Chapter 7 we will consider the various wire antenna systems that can be used when the autotuner is near the rig, feeding the antenna via ladder line feed or a single-wire feed.

MY THIRD CHOICE

My next choice is to use the same physical configuration as the second choice, but with a manual antenna tuner instead of an autotuner. A manual tuner in this configuration will provide a good match, just as the autotuner will. The disadvantage, as compared to my second choice, is that you have to take time to twiddle the knobs manually and read the meters while tuning for the best match. The advantage is that manual tuners are much less expensive than autotuners.

When using a manual antenna tuner with the same antenna every time, you can make up a cheat sheet in tabular form to help you return to the same settings for the same frequency quickly. When using a cheat sheet like that, all you need do is a minor tweaking of the controls, rather than a complete retuning, when you return to a previously used frequency.

There are many manufacturers of manual antenna tuners. You can find many of them listed in the advertisements in *QST*. Don't try to get by with the least expensive tuner you can find. It's better to spend a little more money to get a quality tuner, and one with a wider tuning range. Also, you can check with your ham friends to find out which tuners they have used, and how satisfied they were with the performance of those tuners.

160	2.0	High-Z 2	135
80	8.5	High-Z 2	156
75	2.0	High-Z 2	156
40	8.5	High-Z 1	57
30	9.5	Low-Z 1	20
20	3.5	Low-Z 1	13
17	2.5	Low-Z 1	52
15	1.0	Low-Z 1	7
12	1.0	Low-Z 1	0
10	0.5	Low-Z 1	8

Figure 3 — You can make up a "cheat sheet" like this for your antenna tuner settings. The cheat sheet helps you return to the same frequency with a minimum of on-the-air tune-up time.

Figure 4 — The MFJ-941E Versa Tuner II is an example of a manual antenna tuner for the HF bands and 160 meters.
(Photo courtesy of MFJ Enterprises.)

MY FOURTH CHOICE

The next antenna I would consider using is the *screwdriver* mobile antenna, named because it uses a motor from a handheld screwdriver. For a quick history and description of the screwdriver antenna, originally developed by Don Johnson, W6AAQ, check out **www.antennex.com/preview/Folder01/ screwdrv/screwdrv.htm**.

Once the idea of the screwdriver antenna caught on, several companies started offering commercial versions of the antenna. The screwdriver antennas I'll use as examples here are from the Tarheel line, which you can check out at **www.tarheelantennas.com/mobile_antennas**.

Tarheel makes several variations on their screwdriver antennas. Some tune from 3.5 to 54 MHz, and others from 1.7 to 29 MHz. Some are rated for 200 W PEP, some for 400 W PEP, and others for 1500 W PEP. You can choose the model to suit your station equipment and operating habits.

Screwdriver antennas can be mounted easily in attic spaces. If there isn't enough vertical height in your attic space to accommodate the antenna, it's possible to turn the top whip away from the vertical immediately above the loading coil.

When mounting a screwdriver antenna for fixed-station use, you will need to provide a system of radials to act as the antenna's counterpoise. It's usually adequate to have one quarter-wavelength radial for each frequency band you expect to use.

If you don't have enough horizontal space to stretch out radials for the lower bands (such as on 75 meters, where a quarter-wavelength radial is about 61 feet, or on 40 meters, where a quarter-wavelength radial is about 33 feet), you can always fit the radial into the space that's available. You can run the radial away from the base of the screwdriver in a straight line as far as possible, and then bend

Figure 5 — KC3VO's Tarheel screwdriver antenna mounted on his Gold Wing motorcycle. (For details on KC3VO's motorcycle superstation, see page 4.18 of the ARRL publication Amateur Radio on the Move.)

it or even zigzag it to make it fit into your available space. High RF voltages may exist at the end of radials, so place them carefully to avoid nearby metallic objects and people or pets.

You can use either a manual antenna tuner or an autotuner arrangement with a screwdriver antenna. With a manual tuner, you watch a SWR meter in the coaxial line as you tune the antenna, tuning for minimum SWR (which typically runs about 1.1:1). With the autotuner system, you depress a switch and the antenna tunes itself to resonance by sensing the SWR and stopping at the minimum value. Again, the autotuner system is easier to use, but it costs a bit more. Tarheel offers a turns counter that they call a "programmable counter" for their screwdriver antennas. The unit has ten memories for returning to the same inductance value, making it easy to return to as many as ten specific frequencies at the push of a button.

MY FIFTH CHOICE

My next low-profile antenna choice would be to try to fit a half-wave dipole or other conventional coax-fed wire antenna into an attic space. As with the quarter-wave radials for the screwdriver antenna, the antenna can be bent to fit into the available space. You should, however, avoid doubling the antenna back on itself. If you double the antenna back on itself, the RF currents that flow in the two adjacent wires are out of phase with each other. That tends to cancel some of the overall radiation from the antenna. Rather than doubling the antenna back on itself, try to bend the antenna at a right angle when you run out of straight-line space.

For multi-band use, you can feed a dipole (that is a half-wavelength long at the lowest frequency band you intend to use it on) with ladder line and an antenna tuner that has balanced output.

Most commercially available antenna tuners are single-ended devices, with coaxial input and either coaxial output or two output terminals for balanced feed lines (such as ladder line or twin lead). This is accomplished inside the tuner box by having a balun to convert the single-ended output from the tuner circuit into a balanced output to use with a balanced feed line. Often, the impedance-matching requirements placed on the antenna tuner will cause the balun itself to operate outside its design impedance, causing excessive RF loss.

To avoid the balun problem, the use of balanced antenna tuners has become more common. For a good overview that compares unbalanced and balanced tuners circuits, see the Product Review by Joel Hallas, W1ZR, in the September 2004 issue of *QST*, "A New Generation of Balanced Antenna Tuners." See also the earlier articles referenced in that Product Review.

You can also use a single coaxial line to feed more than one dipole connected together at the center feed point — the *parallel dipoles* arrangement that was first described by H. J. Berg, W3KPO, in his article

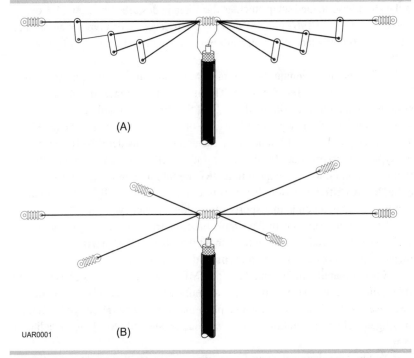

(A)

(B)

UAR0001

Figure 6 — At A, the conventional parallel-dipole arrangement (viewed from the side), with insulators for the shorter dipoles attached to the next-longer dipole. At B, a possible attic-mounted parallel dipole (viewed from above), with the dipoles fanned out horizontally to fit within the available space.

"Multiband Operation with Paralleled Dipoles," in the July 1956 issue of *QST*. If you use multiple dipoles, it's best to run them away from the center insulator at slightly different angles and keep them fairly widely separated, to minimize mutual coupling (and detuning) between or among the dipoles.

One simple antenna that will cover three bands — 40, 20, and 15 meters — with coaxial feed can be made by using two parallel dipoles, one cut for 7 MHz and the other cut for 14 MHz. On 15 meters, the 40-meter dipole will accept power from the feed line and operate as a 3/2-wavelength dipole. If you try this antenna, you may need to experiment with the length of the 40-meter dipole to reach a compromise length that will result in a low SWR on both 40 and 15 meters.

MY SIXTH CHOICE

One very simplistic antenna that is low in efficiency, but which has produced good enough results to get a lot of CC&R-inhibited hams on the air is the short base-loaded whip, such as the antenna described by Lew McCoy, W1ICP, in the June 1967 issue of *QST*. The efficiency of this type of antenna is quite low. But let's face it, if you can't manage anything better, it will get you on the air.

This type of antenna uses a simplistic antenna tuner feeding a short whip antenna that is only a few feet long. The antenna tuner tunes the antenna system to match the coaxial feed line, and the short whip radiates the RF — but with low efficiency — perhaps 5% or lower. Before you get excited about having only 5% efficiency with the antenna, consider this: If you run the typical transceiver at 100 W output, the 5% efficiency means you will radiate a signal that's equivalent to a QRP signal into a good antenna (such as a dipole). An efficiency of only 5% is tolerable, and you will be able to make contacts with an antenna that sports such a low efficiency.

Commercial versions of the simple base-loaded whip can be purchased, such as the MFJ-1622. You can read about the MFJ-1622 at **www. mfjenterprises.com/products.php?prodid=MFJ-1622**.

Several manufacturers, in addition to MFJ, market this type of base-loaded short whip antenna. Also, you can make your own. If you decide to roll your own, use a large-diameter coil with a pitch that's large enough for you to attach an alligator clip to a point on the coil without accidentally touching the adjacent turn.

You may be skeptical about 5% antenna efficiency. You might not believe you can work out of your CC&R backyard with such an inefficient antenna. Let me give you an example of what can be done with a little whip: Chuck Joseph, N5JED, now living in San Angelo, Texas, worked WAS, WAC and DXCC while running 100 W to an MFJ-1621 windowsill whip (the predecessor of the MFJ-1622). Chuck's MFJ-1621 was mounted indoors at his former home in

Figure 7 — The MFJ-1622 is a small and simple antenna that the low-profile ham can use to make good contacts from 40 through 2 meters. Despite the inherent low efficiency of any antenna of this type, whether homebrewed or commercially made, you might be surprised at how well it works. (Photo courtesy of MFJ Enterprises.)

Figure 8 — N5JED worked WAS, WAC, and DXCC while running 100 W to an MFJ-1621 mounted indoors on the second floor of a wood-frame house.

Mount Clements, Michigan. His antenna was on the second floor of his wood frame house.

If Chuck can do that well, you can sure try to match his accomplishments!

MY SEVENTH CHOICE

My seventh choice (and the last one we'll discuss here) is the small-diameter loop. My long-time friend Rod Newkirk, W9BRD/VA3ZBB, used one in his Chicago home for many years with considerable success.

Chapter 13 gives details of Rod's approach to compact loops, and how you can design your own. Again, commercial compact loop designs are available from various manufacturers.

Here are two examples of compact loops. The MFJ-1786 is a 36-inch-diameter loop antenna that provides continuous coverage from 10 through 30 MHz. The MFJ-1788 is a loop of similar size that covers 7 through 21 MHz.

For complete information on those MFJ compact loops, go to **www.mfjenterprises.com/ products.php?catid=3&subcat =hfloop**.

Figure 9 — The MFJ-1786 compact loop and its control box. This is another antenna that the low-profile ham can use indoors, if necessary.

5

Feed Lines for HF Antennas

Before you start planning what kind of antenna you will put up, you should first consider what kind of feed line you will use — or perhaps *will be able* to use. In most cases, the feed line will have to be as nearly invisible as possible, just as the antenna itself must be.

There are three main types of feed line that can be used with low-profile antennas: coaxial cable, ladder line (or homemade open-wire line), and single-wire feed. Here are some thoughts on the use of each type of feed line with low-profile antennas.

COAXIAL CABLE FEED

Inexpensive coaxial cable was developed for use during World War II, when methods were perfected to weave the cable's outside braid at low production cost. Following WW II, millions of feet of military surplus coaxial cable flooded into the surplus radio marketplace at giveaway prices. Many hams, on the lookout for a bargain, switched over to using coax feed lines for their antennas. They got into the habit of using coaxial cable feed lines, and that habit persists to this day.

The main advantage of using coax to feed a low-profile antenna is that it can be buried, or tucked under the eaves of your home to keep it out of sight. With a CC&R problem you can't hang it out in plain sight, as would be the case of using coax to center-feed a dipole hanging between two trees without a CC&R problem.

If you bury coax, it's best to run it inside inexpensive, thin-wall plastic garden hose to give it some protection from moisture in the ground. Even at that, the coax will deteriorate more quickly than would coax run above ground. So be prepared to replace buried coax more often than you would have to replace aboveground coax.

For low-profile antennas, middleweight coax (such as RG-58, RG-59, and RG-8X) is a good choice. Larger coax (such as RG-8 and RG-9) is more difficult to keep out of sight. You could also use much smaller coax, such as RG-174, but it would be more susceptible to mechanical damage. It would also give higher feed line loss and cannot handle much power, but the difference in the feed line loss between RG-174 and RG-8X isn't always enough to worry about at frequencies in the HF spectrum. You may have a situation where RG-174 can serve you adequately, so don't rule it out. As in all aspects of low-profile antennas, be prepared to do anything necessary to put up an antenna — the results might be less than optimum, but they will get you on the air.

Coax feed lines have considerable advantages, but they can sometimes be the limiting factor in the usable bandwidth of an antenna (ie, the bandwidth over which the SWR is reasonably low).

Although the mindset of most hams is that the antenna's SWR must be at or near 1:1, you can do very well when using antennas at higher SWR readings. My own rule of thumb is that the SWR should be no higher than 3:1. However, some modern rigs will start to reduce their output at SWR values of 2:1, to keep from damaging components in the rig's output stage. Check the equipment manual for your rig to determine the SWR at which the rig starts to provide reduced output, and then aim to have a SWR no higher than that value.

Now, let's get back to your antenna's SWR readings across a given frequency band. Your antenna may not have a usable bandwidth wide enough to cover all the frequencies you want to use within a single amateur band, especially on the lower HF bands. This is especially true of shorter antennas that are less than a half-wavelength long. If you use a *Marconi* antenna, with a vertical or semi-vertical element working against a counterpoise or a good earth ground, you can aim for a minimum antenna length of about a quarter-wavelength.

If you have a usable bandwidth that's too narrow to include all your favorite operating frequencies within a given ham band, you could use an antenna tuner (either an external tuner between the rig and the coaxial feed line or an internal tuner, if your rig has one) to reduce the SWR at the transmitter end of the coax feed line. This will make your transmitter happy.

Another possibility that can be used to increase your antenna's useable bandwidth is to install an autotuner at the antenna's feed point. As mentioned elsewhere, the SGC tuners present a good choice for this application. You can build your own remote antenna tuner, but such a project will require good mechanical design skills and building techniques, in addition to good electrical design and construction skills. Most hams nowadays would, instead, opt to buy a ready-made unit.

BALANCED FEED LINES

The term *balanced feed line* means a two-conductor feed line with side-by-side conductors. There are several common types of balanced line that hams use. The old-fashioned (and now uncommon version) is *open-wire line*. Two conductors (usually bare, uninsulated wires) that run alongside each other remain generally parallel, held at the same conductor-to-conductor spacing with insulated spacers along the length of the line. Open-wire feed lines were very common 50 to 70 years ago, but they don't maintain the same spacing between conductors at every point along the line. When the feed line blows in the wind, the spacing between conductors can vary and cause your SWR to fluctuate. In worse cases, the feed line can twist and, if the conductors aren't insulated, they can short out.

In more recent years, following the electrical idea of open-wire line, manufacturers started making *ladder line* (sometimes called by the more accurate name, *window line*). Figure 10 shows examples of ladder line. As you can see, the insulating webbing of the ladder line holds the spacing between the line's two conductors virtually constant.

Figure 10 — Ladder line (also called "window line") is a balanced feed line. Here are two types of ladder line, one with a single conductor for each side of the line, and the other with stranded conductors. In both cases, the conductors are usually Copperweld wire (copper-plated steel wire).

Ladder line is made with two different types of conductors. One type uses a single conductor for each side of the line, while the other uses a stranded conductor (a small bundle of conductors) for each side of the line. The wires in each type of ladder line are usually copper-clad steel — the steel core provides mechanical strength, while most of the RF energy flows on the copper surface of the conductor (the so-called *skin effect*).

Stranded-conductor ladder line, generally speaking, is preferable. The main reason for that is that the solid-conductor line is less likely to break because of the repeated flexing that the antenna-to-feed-line connection experiences. However, when using ladder line for low-profile antennas, the ladder line must remain out of sight (perhaps under the eave of the house, but away from metallic objects) and firmly attached to the house. Therefore, it's less likely that there will be any flexing of the antenna-to-feed-line connection, so either type of conductor would be acceptable.

If you use ladder line as your feed line, you will need to have an antenna tuner with balanced output. Almost all antenna tuners that have balanced output are actually unbalanced antenna tuners with a balun at the tuner's output to convert the unbalanced output to a balanced output.

There are some true balanced antenna tuners on the market today. Those tuners are symmetrical tuners, rather than a single-ended tuner with a balun at its output. For a good look at some of those tuners, as well as a technical discussion of their operation, see the Product Review, "A New Generation of Balanced Antenna Tuners (MFJ-974(H); Palstar AT4K; Palstar AT1500BAL; E.F. Johnson Matchbox)" in Sep 2004 *QST*.

When choosing a commercially built antenna tuner, pay careful attention to the manufacturer's specification of the impedance range over which it will tune. When dealing with the kind of antennas you will be forced to use in a low-profile situation, you should aim for an antenna tuner that will tune over a range of about 30 to 2000 Ω. You may at times encounter some antenna/frequency combinations that will require a tuner that will cover a range that wide.

Another reminder about the autotuners that are built into some of the commercially available transceivers: If you check the specifications in your owner's manual, you will find that the impedance range over which the autotuner can tune is pretty narrow — perhaps covering 40 to 100 Ω. Those autotuners are designed to tune out only small mismatches in impedance when using coaxial cable as the feed line, and will not cover the large excursions in feed line impedance you may experience when using low-profile antennas. We'll discuss antenna tuners at greater length in Chapter 12.

Getting back to the discussion of balanced transmission lines, a lot of hams use TV type *twin lead* as a balanced transmission line, rather than ladder line. Twin lead has the advantage in low-profile installations because it is smaller and therefore easier to conceal. Electrically, it works essentially the same way ladder line does.

Woods Industries makes twin lead with clear insulation (their model 341), which is even better for low-profile hamming. Their clear twin lead can typically be found in 500-foot rolls for $69. Aubuchon Hardware sells the Woods clear twin lead at their online store; see **electrical.hardwarestore. com/15-54-tv-wire-cable/500"-flat-lead-antenna-wire-clear--628422.aspx**.

SINGLE-WIRE FEED LINES

Most hams who use single-wire feed simply end-feed the antenna at the nearest end, continuing the wire from the antenna into the ham shack, where it is connected to the antenna tuner. It's a simple way to feed the antenna and it can work okay, but it has some disadvantages. The main disadvantage is that relatively high RF voltages may be present on the single-wire feed line that's coming in your window and going over to the operating position. There is the possibility of getting RF burns from the line (if it's not insulated), of excessive

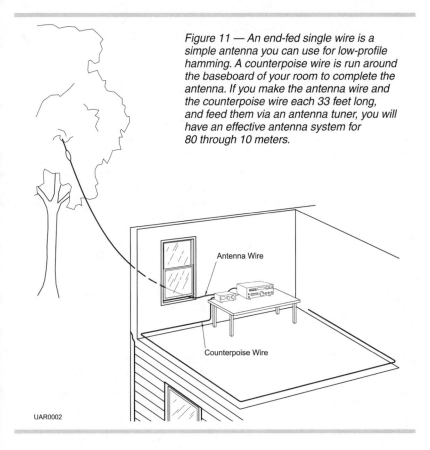

Figure 11 — An end-fed single wire is a simple antenna you can use for low-profile hamming. A counterpoise wire is run around the baseboard of your room to complete the antenna. If you make the antenna wire and the counterpoise wire each 33 feet long, and feed them via an antenna tuner, you will have an effective antenna system for 80 through 10 meters.

Antenna Wire

Counterpoise Wire

UAR0002

exposure to RF energy and also of causing interference to consumer-electronic devices in your home. Therefore, feeding your antenna with a single wire should be your last choice.

The main advantage of single-wire feed is that it is less visible than the other types of feed lines. If the wire gauge used for the antenna and feed line is small enough, it can be run in the open without attracting attention. The single-wire feed line should be kept clear of other objects as much as possible — especially metallic objects.

Old-timers will remember not only the Windom antenna that was fed at an off-center point with 300 Ω TV twin lead, but also the original Windom antenna that was fed with a single wire connected to the antenna flat top at an off-center point. The good old Windom with single-wire feed is still a capable antenna.

This illustrates the advantage of doing a lot of reading on the subject of antennas when you start working on your low-profile antenna, and of reading older literature as well as recent literature. If you know about the less-well-known antenna designs and the less familiar antennas hams have used in the past, you will be better equipped to decide on an antenna design that will fit your situation.

6

Some Simplistic Antenna Theory

This chapter will give you a small start toward understanding antenna theory and its associated formulas and graphs. After you see how easy it is to learn to understand those areas, you might become curious enough to start digging deeper into antenna theory.

A LITTLE MATH

Okay, here's the first math: An antenna that is a half-wavelength long can be a very effective antenna on the HF bands. It's good to get a feel for the size of a half-wavelength antenna, so you can try to get that much wire in the air. You should learn this number:

468.

Memorize it.

(Pause…)

Now that wasn't so hard, was it? If, at first, you have trouble remembering this number, you can note it on your Redneck Palm Pilot.

So what can you do with the number 468? If you divide 468 by the frequency (in megahertz) for which you want a half-wave antenna, you will determine the length (in feet) of a half-wavelength antenna. For example, to determine the length (L) for a half-wavelength antenna for 7.090 MHz:

L = 468/7.09 = 66.0 feet.

You are on your way to learning about antennas! Doesn't that make you feel good? Note that this formula is for a half-wavelength *antenna* — a half wavelength in *free space* is different. But for our present purposes, we won't think about that. It might make our brains hurt.

With a half-wavelength 40-meter antenna at a fair height above ground, say 30 feet or so, you can get out

Figure 12 — The Redneck Palm Pilot can be used for recording data for later use. Here, it is shown storing our Magic Number for Half-Wavelength Antennas: 468. (Hint: Use a permanent marker for long-term storage, or a washable marker for short-term storage.)

pretty well, make plenty of contacts and enjoy your HF operating. Now that we know the length of a half-wavelength antenna for 7.090 MHz, what else do we need to know?

Oh, no — here come those pesky graphs with some mysterious meaning! Take heart and have confidence in yourself: You have already mastered an important math equation. Can graphs be so difficult to learn? I think not.

A SIMPLE GRAPH

There's one kind of antenna diagram you should learn to understand to get a feel for what performance to expect from an antenna. That graph shows the current distribution along the length of a half-wavelength antenna. Check out Figure 13.

The straight heavy line represents a half-wavelength antenna, as viewed from the side. The lighter-weight sinusoidal line shows the current distribution along the half-wavelength antenna. The radiation from the various points along the antenna's length is directly related to the RF current at those points. The greater the current at any given spot along the antenna's length, the more it is radiating from that point.

Look at the heights of the sinusoidal line above the baseline at points as you move from one end of the antenna toward the middle. Near the end, there's not much height, which means there's not a lot of current flowing at that point on the antenna and not much radiation taking place. Farther toward the middle

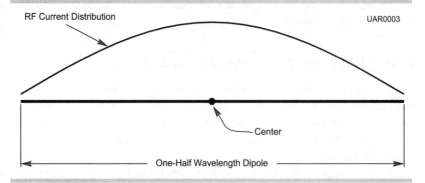

Figure 13 — The current distribution along a half-wavelength dipole antenna. The heavy straight line represents the dipole, and the lighter sinusoidal line shows the current distribution.

of the antenna there's more current, and therefore, more radiation. The peak antenna current is flowing at the middle of the antenna, and that's where the maximum radiation from the antenna is taking place. And it's good to get the high-current portion of your antenna as high and as much in the clear as possible.

So far, we've only been taking a two-dimensional look at a vertical slice of the radiation pattern. It's time to think in three dimensions. Imagine rotating that sinusoidal line (that represents the current flow) around the antenna, forming a sinusoidally shaped, solid, current-flow diagram. Now imagine yourself looking at the end of the antenna. The radiation takes place in all directions around the dot that represents the end view of the antenna — up, down, left, right, and everything in between. If the antenna were in free space, so there were no reflections, the radiation would be equal in all vertical azimuths.

But real antennas don't exist in free space. They are near objects that affect their radiation pattern — notably, the ground itself. When you start reading more about antenna theory, you will see how ground reflections of the transmitted wave front will affect the overall radiation, and the height of the dipole above the ground will affect the vertical *takeoff angle* of the radiation from the antenna. We won't get into that any more here. I point it out to you to pique your interest in learning more about how antennas work.

Our conclusion? The "getting-outest" part of the half-wavelength antenna is at the middle of the antenna, where maximum RF current is flowing. Therefore, that is the part of the antenna that should be placing the highest above ground and the most in the clear. Keep that in mind when you're putting up a low-profile antenna.

Note that no current is flowing at the ends of the antenna, so the graph of antenna current goes down to zero. We conclude that the radiation from the ends of the half-wavelength antenna is low.

A SLIGHTLY MORE COMPLEX GRAPH

Let's look at another aspect of the radiation pattern from a half-wavelength dipole. Figure 14 shows the three-dimensional radiation pattern from a half-wavelength dipole (again, the antenna is in free space) shows that the maximum radiation is broadside to the antenna, in a pattern shaped like a Krispy Kreme doughnut.

I have seen more than one ham work hard to orient their dipoles within 5° of the desired direction, so they get maximum radiation in a desired direction. Hold on! Look at the figure again. Note that the angles (away from broadside) that the antenna is 3 dB down from its maximum radiation are at the ±45° points away from broadside to the antenna.

It's generally accepted that it's hard to hear a 3-dB difference in signal strength by ear. So the antenna pattern actually covers 180° (90° on each side of the antenna, as you look down on it from above) with essentially the maximum radiation.

But, wait! There's more! (Oops! Pardon me for sounding like a TV infomercial there.) Look at the points where the radiated signal is 12 dB down from maximum (broadside) radiation. An S unit is generally considered to be 6 dB. Dang! You are radiating a signal that's no more than 2 S units down from the maximum radiation over 300° of azimuth! Not bad!

The radiation pattern of the half-wave dipole suggests that there is a null off the ends of the antenna. Maybe in free space, in the Land of Theoretical

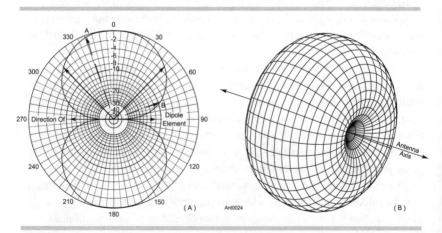

Figure 14 — Directive diagram of a half-wavelength dipole in free space: At A, a view of the dipole from the side. At B, three-dimensional view of the dipole's doughnut-shaped radiation pattern.

Results. But you will get a fair amount of radiation even directly off the ends of the dipole. The radiation from the ends might be vertically polarized instead of horizontally polarized, but vertical polarization can make contacts, too.

In summary, if you put up an antenna that's about a half-wavelength long, it will radiate quite well in all directions. This serves as a good example of your needing to understand not only the theory behind antennas, but also the details of the practical application of that body of theory to actual antennas.

One time I attended an antenna talk at a hamfest where the presenting ham had done antenna modeling to design an array of vertical antennas to use as a directional array. He commented that he had made one small change that resulted (according to the computer software) in a 0.1 dB increase in the radiated signal. (Remember, you can't hear a 3-dB difference!)

Then he showed some color slides of the antennas. They were so far away from good engineering practice (because of his having to fit the antennas into the landscaping and among the trees of his yard) that there would be no comparison of theory with reality. Forget that 0.1 dB advantage, OM!

This true story illustrates W1AB Theorem Number Two: "Don't let your computer make you think you are smarter than you really are." Oh, you want to know about W1AB Theorem Number One? "Don't buy a piece of electronic equipment that's smarter than you are." Theorem Number One is applicable to a great number of people who buy computers, video recorders, cellular telephones and state-of-the-art amateur transceivers.

But let's get back to looking at our dipole. If you cut that dipole to be about 66 feet long, so it's a half-wavelength on 40 meters, and feed 15-meter RF to it, what happens? It will accept RF, even with coaxial cable feed, because the center feed point will be a high-current (and, therefore, low-impedance) point. The SWR might be slightly higher than it is on 40 meters,

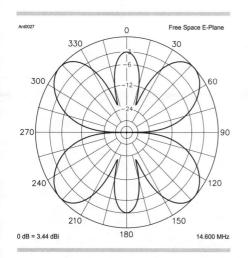

Figure 15 — Radiation pattern for a 3/2-wavelength dipole in free space. Note that the pattern has broken up into various lobes, as compared with the simple lobe structure of the half-wavelength dipole.

but the antenna will work on 15 meters.

However, the plane view of the radiation pattern, instead of having only two broad lobes, will have four major lobes in a sort of four-leaf-clover pattern plus two minor lobes broadside to the antenna. Figure 15 shows the free-space radiation pattern of a ½-wavelength pattern.

Again, even though the radiation pattern shows theoretical nulls, we know that the real world will prevail, and that you will still get radiation in virtually every horizontal azimuth from the antenna.

This is just scratching the surface of antenna theory. I hope I didn't make your brain hurt. But you see how easy it is to learn to understand antenna formulas and graphs, by taking it one small bite at a time. You can read the ham antenna literature, such as *The ARRL Antenna Book*, to better understand antennas. You can read the antenna articles in *QST* and other ham magazines. Find an experienced and technologically competent ham near you who can provide some mentoring for you. The better you understand antenna theory, the better prepared you are to figure out the best antenna possibilities that are within your own low-profile limitations.

Furthermore, I encourage you to learn more about the practical aspects of antennas by building various types of antennas and experimenting with them. Find out what works best, and then try to figure out why. Do that and you will be better prepared for your next antenna project.

7

Low-Profile HF Wire Antenna— Indoors and Outdoors

There are plenty of indoor and outdoor antenna possibilities available to the low-profile amateur operator. Let's first take a quick overview of current usage on some of the more popular HF ham bands, so you can consider what bands you might want to try operating on.

HF BAND USAGE, 80 THROUGH 20 METERS

The 80-meter CW subband, over the past few decades, hasn't been used nearly as much by US hams as it was in the 1960s and '70s. Furthermore, from the viewpoint of the low-profile ham, effective antennas for 80 meters need to be twice as long as comparable antennas for 40 meters. Putting those two considerations together brings me to the conclusion that the low-profile ham could do better by concentrating his or her efforts on other HF bands.

However, don't exclude 80 CW from your thinking. It can be a productive band, especially during the long winter nights when the static levels are low and the skip is long. Therefore, you should give 80 CW a try sometime.

The 75-meter SSB subband is very popular and heavily used,

especially during the evening and in the morning hours just before commuting time begins. During daylight hours, you have limited sky-wave propagation, so most of your contacts will be with relatively nearby hams.

Many of the stations on 75 SSB run high power, many use large and effective antennas, and many use *both* high power and large antennas. That means that, especially during the nighttime peak hours, your competitors for spectrum space will be able to outgun you severely when you're using indoor antennas or other low-profile antennas and modest power. You might find a clear frequency and begin a contact on it, but the probability is very high that another station or group of stations — likely stations with big signals — will pretty quickly move on or near your frequency and cause your contact to be lost. However, if you want to try 75-meter SSB from your low-profile station, do have a go at it, but be prepared to have a fairly low success rate.

The 40-meter CW subband is a good band for overall hamming. The band generally has propagation paths open to somewhere 24 hours per day. During daylight hours you can work stations via short-skip paths of a few hundred miles, and by night you can work long-haul DX virtually anywhere in the world — even with low-profile antennas. Therefore, 40 meters is a good choice for the low-profile ham.

The longer propagation paths on 40 CW form during hours of darkness. Beginning with daylight at your location, let's take a quick look at how the band evolves as darkness falls. You will begin to hear stations via long skip who are to the east of your location — when darkness has fallen in *their* locations but while it's still daylight at *your* location.

Later, when darkness has fallen in your own location as well, you'll be able to make solid contact with those stations to your east. Then later you will be able to make contacts via long skip with stations to your west, and to your north and south. Later yet, as propagation continues to improve, you will begin to hear stations at longer distances, either via improved single-hop propagation or via double-hop propagation.

When the areas of darkness on the Earth's surface are "just so," you can often work DX via the *long path* — that is, by the longer great-circle path between you and the DX station, rather than the usual shorter great-circle path. Generally speaking, for those long-path contacts you must have darkness along the entire length of the long path. To me, long-path DXing is very interesting. Making a contact over a path length that is more than halfway around the world is fascinating.

One very interesting phenomenon in HF radio propagation is the effect called *gray-line propagation*. If both you and a distant station are in partial darkness, either just before or after sunset, or just before or after dawn, your signals will be propagated with enhanced signal levels along that gray-line path.

For a good moving-image graphic that shows where the gray lines

One of My Best DX Contacts Ever — via the Gray Line

And I am not talking about the tour-bus company.... In the 1990s, when I was living and working in Maryland, I used to commute to work on the Honda Gold Wing motorcycle I owned at the time. One morning (August 3, 1993) I woke up about 4:00 am and couldn't get back to sleep, so I decided to go to work early.

My drive to work took 45 minutes, during which I usually chatted with friends on the low end of 40-meter CW from the cycle. My station was a Kenwood TS-140, MFJ-484 grandmaster memory keyer, and a Hustler whip with a two-band top on it for 40 and 20 meters. That particular morning, I was riding to work about two hours earlier than usual, and none of the morning crowd was on the air yet.

I tuned to the low end of 40 CW to see what DX might be up and about. The first station I heard was Yuri, 4K1F, on King George Island in Antarctica, with a solid 599 signal. I called Yuri, and he came back to my first call, giving me a 599 report (at 1018 UTC, 0618 EDT). We finished our contact, and I first assumed that he was giving everyone 599 reports. I continued to listen, and heard him give other stations lower signal reports. Wow — I must have really been strong!

After I got to work, I finally realized that I had worked Yuri via the gray-line path. That's why we both had such big signals.

Figure 16 — K3KMO motorcycle mobile, on the move and ready for rag-chews and DX.

currently exist, go to **www.fourmilab.ch/earthview/vplanet.html** and click on "map of the Earth." Try listening at some of your local gray-line times, and you may be surprised at the strong signals you can hear over a long-haul DX path.

This is a just a broad-brush introduction to 40-meter propagation. If the subject tweaks your interest and imagination, you should read more about DXing and propagation on 40 meters, to better understand what might be available to you and your low-profile antennas.

On 40-meter SSB, you can make good short-skip contacts during daylight hours. However, the long-skip propagation during hours of darkness presents a problem, because the 40-meter SSB subband is loaded with high-power international broadcast stations. It's difficult to find a clear frequency because the broadcast stations fill the band (as the CBers used to say, "wall to wall and ten feet tall"). It's possible to make long haul and DX contacts on 40 SSB with only a modest station and antenna, but it takes a fair amount of patience and effort to do it.

As you go higher in the frequency spectrum, the antennas become smaller in inverse proportion to the frequency. That's good news for the low-profile ham. At the lower part of the sunspot cycle, when propagation isn't so hot, the higher-frequency bands may not be very productive for either local or DX contacts. But when we're near the peak of the sunspot cycle, you can make good DX contacts with your low-profile antennas, both during daylight hours and at night.

Many low profile and QRP (low power) hams use 30-meter CW to great advantage. It's an excellent band both for rag-chewing and for DXing.

When the 20-meter band is open, it's a good band for long-haul DXing. However, on both CW and SSB (especially the latter), you will bump heads with some serious competition — there are a lot of hams with high power and big antennas.

A great many hams have enjoyed working the so-called "WARC bands," the newer ham bands (12, 17 and 30 meters) that are not in the usual harmonically related group.

There's lots of HF spectrum space available for your use. Set up your station and antennas so you can operate on all the HF bands. Try the various bands at various times of day and night, to learn what you can expect from each of them. Then when you sit down for an operating session, you will know from your own experience (with your own antennas) what you might expect to find in the way of other stations to contact.

And now, let's consider some antenna possibilities.

INDOOR ANTENNAS

Some of the main considerations for an indoor wire antenna include the following: The antenna should work well, it should be approved by the

My Two Years in an Apartment

In the early 1970s, my wife Maggie and I lived in a very nice apartment complex in Wheaton, Maryland (a northern suburb of Washington, DC). The apartments were built as a variation on garden apartments. Separate apartment modules were connected with a common wall of concrete-block construction. Each module contained six apartments (two on each of three floors) with each module slightly offset from one to the next, to avoid the long, continuous look of the typical garden apartment building. Each module had a square, four-sided peaked roof, which gave the apartments a slightly Scandinavian look. It was a nice place to live.

Did I tell you that the apartment complex did *not* allow any outside antennas? I didn't think I needed to….

My favorite all-band HF wire antenna at that time was the 130-foot dipole, center-fed with open-wire line via an antenna tuner that had balanced output. Our apartment was on the third floor, and there was a hatch in our apartment that gave me access to the large attic that topped our apartment module. I installed a 130-foot dipole in the attic by placing its center at the peak of the roof, about 35 feet above the ground. I ran the dipole's two 65-foot wires down the inside of the roof to two of the roof's corners, and then ran the remainder of each side of the dipole along the base of the roofline. Viewed from above, the antenna made a Z shape. Looking at it from the side, it was an inverted-V dipole with the ends bent sideways at an angle of about 135°.

The antenna worked quite well, and I made plenty of good contacts with it on the various HF bands. The downside of operating from that apartment was because of its community TV system. An antenna on the roof picked up the broadcast signals, which were then run via coax to the basement of the apartment, where an amplifier and distribution box was located. That box contained a wideband, high-gain amplifier that fed a six-way splitter to feed the TV output jack in each apartment. It was a recipe for TVI!

Sure enough, I started getting TVI complaints. I curtailed my operating hours, to avoid the times that people would be watching TV, while I worked out a solution. I contacted the vendor that supplied and maintained the apartment's TV system and explained the problem. To their credit, they soon installed a replacement amplifier that had a high-pass filter on its input. The new amplifier filtered out my HF signals quite well, so they weren't delivered to my five neighbors' TV sets. Problem solved!

---

ham's spouse (*very* important), and it should be mounted in an out-of-the-way manner, to avoid becoming a tripping hazard. In a home of average size, to include both apartments (or condos) and single-family houses, there is plenty of room for antennas for the HF bands.

You can put up an antenna inside a room in your home. You can run a wire around the top of the walls in your ham shack, and feed it either in its center as a bent dipole, or connect the ends of the wire to make a loop and feed it at a corner or in the middle of one side of the rectangle.

You can feed a random-length wire at one end. If you do that, a good, low-impedance RF ground is necessary. If you have trouble getting a good RF ground, you can use a single counterpoise wire, run

Figure 17 — The MFJ-934 has both an antenna tuner and a ground tuner (to tune a counterpoise) in the same box, a good combination to help the low-profile ham cope with his compromise antenna systems.

around the baseboard of the room your ham shack is in. You can also use an artificial ground such as the MFJ-931, or a combination tuner such as the MFJ-934. The MFJ-934 combines an antenna tuner and an artificial ground in the same equipment cabinet, which is convenient.

If you have access to your attic, there is usually plenty of attic space for HF wire antennas. Appendix B describes some HF attic antennas that W4KVS used with good success.

The bottom line is that you can try almost any kind of indoor wire as a low-profile antenna. What you first try may or may not work well. If it does work well, enjoy making contacts to try it out. If it does *not* work well, then try something else.

OUTDOOR ANTENNAS

Outdoor antennas are a good news, bad news area for the low-profile ham. The good news is that they usually work better than indoor antennas. The bad news is that people can see them.

As was mentioned in Chapter 3, low-profile hams can use outdoor antennas (1) that are virtually invisible, (2) that can be put outdoors under cover of darkness, and brought back inside before the dawn's early light, or (3) that are disguised as objects that *are* allowed (such as flagpoles). We'll look at disguised antennas in Chapter 10, and in this chapter we'll consider items 1 and 2 above.

INVISIBLE ANTENNAS

There are several ways to make a wire HF antenna virtually invisible, so the casual passersby won't notice it. One way to make the antenna virtually invisible is to run the wire around the outside of your home, keeping it snug up against the wall and hiding it as much as possible behind nonmetallic parts of the structure. In Appendix A, you can read how NØADL accomplished that.

Another way to hide the antenna wire in plain view is to use very small gauge wire — #26 or smaller — and run the antenna as high above the ground as possible. I've done this in many places, and people usually don't notice the antenna. My long-time friend Mike Jaquish, W4WJH, took some of my 26-gauge antenna wire (with brown insulation) home from our Sint Maarten vacation together and used it to make an end-fed 85-foot antenna for all-band use. It was higher at the fed end, going down to a lower support at the far end, so it became a sloper type of antenna (toward the southwest). Mike reports that the sloper definitely favored the southwesterly direction. Even though Mike doesn't have antenna restrictions, he figured there's no use becoming a target of imagined-TVI complaints from his neighbors. Thus he chose the small-gauge wire.

That small-gauge wire served Mike well for about six months but finally came down during a storm with high winds. Mike replaced that wire with a larger #15 wire. Within a few months, a neighbor asked about his new antenna — she had never noticed its predecessor, even though it was in plain view!

Another way to make a wire antenna nearly invisible is described by WB3BJF in Appendix I. He built a multi-wire vertical antenna by attaching the vertical elements directly to the outside of a tree (remember to use the Official Nomenclature, "tee-are-double-e").

Quite a few hams have used aluminum gutters on their homes as antennas, with considerable success. If the sections of the gutter are bonded together extremely well, so their connecting points are low-impedance RF connections, it can be a good antenna. But the various sections of guttering *must* be connected together with low-impedance connections.

If you plan to use your gutters as an HF antenna, you should go along the gutter's length and securely bond mating sections together. Scrape some of the paint off the inside of the gutter near the ends of the mating pieces and either bolt them directly together or install a short bonding strap from one piece to the next.

If, however, you try to use the gutter as an antenna without making sure all its sections are electrically bonded to one another, here's what might happen. The point at which two sections of the aluminum guttering come together can be a high-impedance RF connection. (Hey! It still will hold water, and that's all the gutter installer cares about.) That high-impedance connection can act as a crude semiconductor, in the manner of the old-fashioned cat's-whisker crystal detector (you Young Squirts might have to do some reading about that early device). That semiconducting connection can rectify (detect) your RF signal and then re-radiate your signal across a very wide spectrum, re-using the gutter as *that* signal's antenna — a recipe for potential interference to any electronic device in your immediate vicinity and in your neighbor's homes.

Other hams have used the metal flashing installed at the lower edge of the roof as their antenna. (That flashing is used in the northerly latitudes to prevent ice dams at the edge of the roof, which can then cause the melting snow and ice on your roof to leak into the house.)

NOCTURNAL ANTENNAS (EINE KLEINE NACHTANTEN?)

If you want to go to a little extra trouble, you can use an antenna system that you hide indoors during hours of daylight and deploy outdoors under cover of darkness. The good news is that you have considerably fewer limitations with respect to where, how, and how long you make your antenna.

The *bad* news is that you have to take it in again at the end of each operating session. The *worst* news is that some inclement night you are not going to want to get wet and cold by going out into the bad weather to take in your antenna. Don't *ever* be so tired (or just plain lazy) that you leave your nocturnal antenna out to greet the sunrise.

8

The Killer Antenna ... and Some Variations

THE KILLER ANTENNA

After I moved to New Hampshire in 1998, I put up a multiband HF antenna that I called the "Killer Antenna," and described it in an article in the July 1999 issue of *QST*. The antenna was intended to be a temporary antenna — easy to put together and easy to get into the air — but it worked so well over the following winter that it remains my main antenna to this day.

Note that the Killer Antenna can be easily adapted for low-profile use, by using small-gauge antenna wire and routing the wires and the ladder line in a manner that will hide them.

The original Killer Antenna, still in use, is simply a 40-meter elevated ground-plane antenna that is fed with ladder line through an antenna tuner with balanced output. The vertical wire is 33 feet 3.5 inches long (a quarter-wavelength near the low end of the 40-meter CW band). Three wires of the same length are arranged to form a counterpoise that is hung about 6 feet above the ground and insulated from ground. A reprint of the original article appears as Appendix G in this book.

One advantage of an antenna like the Killer Antenna is that it requires only one support point high in the air, plus three support points that are within arm's reach while standing on the

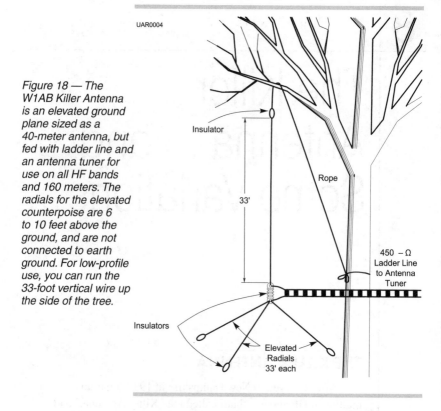

Figure 18 — The W1AB Killer Antenna is an elevated ground plane sized as a 40-meter antenna, but fed with ladder line and an antenna tuner for use on all HF bands and 160 meters. The radials for the elevated counterpoise are 6 to 10 feet above the ground, and are not connected to earth ground. For low-profile use, you can run the 33-foot vertical wire up the side of the tree.

ground. For the original Killer Antenna, I used a slingshot to put a lightweight line over a tree limb about 40 feet above the ground, which I then used to haul up a ³⁄₁₆-inch Dacron rope as the vertical support.

If you feed that antenna configuration with 50 Ω coaxial cable, it would be a dandy single-band 40-meter antenna. However, I wanted to use it on all bands, so I fed it with 450 Ω ladder line through an antenna tuner with balanced output. It loads well on all bands from 160 through 10 meters and it works out like crazy.

I get good signal reports from DX stations on all bands, from 80 through 10 meters. I have often worked DX via the long path on 20 and 40 meters. (I know we're using the long path because either the DX ham tells me he has his beam turned toward the long path, or it's a time of day when the short path is closed.) On 80 meters, I have made comparison checks when working Europeans, and they have always reported that my signal with the Killer Antenna was the same as my signal with my 80-meter dipole. On 160 meters, I can easily work stations within 1000 miles, but working Europe requires good band conditions and operator perseverance on both ends.

AND THEN ... VARIATIONS

After I saw how well the Killer Antenna worked for my home station, I started experimenting with variations of the idea from several portable locations during some of the vacation travels that my wife Maggie and I were making.

My first experiment was a portable antenna from a pair of wires (of random length), with one wire connected to each side of the ladder line. The reason for using wires of random length was to make use of available supports for those wires. I would try to get a total of at least an estimated 100 feet of wire in the air (the sum of the two lengths). That would make the antenna long enough to work well on 80 through 10 meters.

Depending on what supports were available, I might make one of the wires vertical (or nearly so), and the other one generally horizontal (but off the ground). In cases where there weren't any good supports for running one of the wires vertically, I would just run them both horizontally.

Sometimes the antenna would load well on all bands except one. Apparently the impedance at the feed point on that band would be outside the range that the antenna tuner could match.

A little thought gave me an idea. (For ease in following this next discussion, let's call the longer of the two original wires Wire 1, and the shorter, Wire 2.) I put up a third antenna wire (Wire 3) and connected it to the same side of the ladder line that Wire 2 was connected to. I made Wire 3 a considerably different length from Wire 2.

Why? I reasoned that, on a given frequency band, the combination of Wire 1 and Wire 2 had presented an impedance value at the feed point that was outside the range of the antenna tuner. But Wire 1 and Wire 3 would present a different feed point impedance. With luck, that impedance would be within the tuning range of the antenna tuner.

And, as it turned out, my three-wire antenna tuned well on all bands, including the one where it didn't work before. On the band where Wire 1 and Wire 3 were doing the major radiating, you might wonder what Wire 2 was doing — it was just hanging around. On the band where the antenna wouldn't tune earlier, Wire 1 and Wire 3 became the antenna. Wire 2 presented a high impedance; it wasn't taking much antenna current and wasn't doing much radiating. But on other bands, Wire 2 was likely doing more useful work. My philosophy is: if it works, graciously accept it. I don't try to figure it out — it might make my brain hurt!

Now the Killer Antenna idea has evolved into a three-wire antenna, with wires of random and unmeasured length. It's becoming simpler. Let me tell you about some of the antennas I used from various vacation locations, following this idea.

After we retired, my wife Maggie and I have been traveling as much as we can afford (and sometimes more). We travel both within the United States and to overseas destinations. I've assembled a nice little HF station to take along on our travels, built around a Kenwood TS-50.

A small hard-shell, carry-on–size suitcase holds almost all of the portable station's equipment: the Kenwood TS-50, an MFJ-4225MV 25-A switching power supply, an MFJ-934 antenna tuner (with artificial ground tuner), a good old MFJ-484 Grandmaster memory keyer, an Autronic keying paddle, a WW II Navy-knob hand key, connecting cables (power, antenna, and keyer), small headphones, spare fuses for every fuse in the station, basic hand tools, a soldering iron, a small spool of #26 insulated antenna wire, a short length (about 20 feet) of 450-Ω ladder line, and the equipment manual for each piece of equipment. Some hand tools that Transportation Security Agency personnel might opine to be too dangerous for air travel in carry-on bags have to go into checked baggage, with the hope that the baggage won't get lost or seriously delayed (so far, my checked baggage has always arrived on time — knock on wood!).

Some of the places I've enjoyed operating my suitcase-portable station include coastal "Down East" Maine; Nanucket Island, Massachusetts; Cap Pelé, New Brunswick; Hilton Head Island, South Carolina; Aviemore, Scotland; Philipsburg, Sint Maarten and Rockville, Maryland;. I've enjoyed good results with my variations on the Killer Antenna, which have, coincidentally, been of the low-profile variety.

DOWN EAST MAINE

One summer, my wife and I rented a cottage near the seashore in Down East Maine. There were tall trees behind the cottage. I threw a line into a tree and pulled up a wire that was about 70 feet long, and which ran from the ham shack to the tree at about a 45° angle to the horizontal. I ran out two other wires (Wires 2 and 3), and ran them more or less horizontally (but not connected to ground). Those wires were about 35 and 25 feet, respectively. I ran a short piece of ladder line from the antenna tuner to just outside the windowsill and connected the three wires to the ladder line. That antenna loaded well on all bands.

NANTUCKET ISLAND

During a week's vacation in a timeshare cottage on Nantucket Island, I took some telescoping aluminum tubing, which I used to put up a 25-foot vertical (Wire 1). I ran Wires 2 and 3 (of different lengths) along the side of the house, one in each direction from the base of the vertical. Again, I fed the wires with a short piece of ladder line. And, again, the antenna loaded well on all bands. During that weekend I worked quite a bit of good DX with my

TS-50, including BY1PK through a heavy pileup on 20 meters. The antenna obviously got out okay.

NEW BRUNSWICK, CANADA

Still another time, we took a vacation trip to the north coast of New Brunswick, Canada, just across the bay from Prince Edward Island. I was faced with an interesting and challenging antenna situation there — the tallest tree in the cottage's yard was only about 15 feet tall. There wasn't much in the way of antenna supports. The cottage itself wasn't very tall, with the peak of its roof only about 12 feet above ground.

I ran a short length of ladder line from the antenna tuner through the shack window, so the end was only a few inches outside the window. I ran Wire 1 from the ladder line up and over the top of the roof and across the front yard. I tied the end of the wire to a 6-foot pole near the road. I estimated the length of Wire 1 to be about 80 feet. At its highest point, it was about 12 feet above ground, and it then trailed downward to about 6 feet above the ground.

Wire 2 was run from the ladder line into a tree in the back yard. Its average height above ground was about 8 feet, and it was about 30 feet long. Wire 3 ran along the back wall of the cottage and then to another small tree in the side yard. That wire was about 20 feet long and averaged 4 feet above ground.

All in all, that antenna wasn't very impressive to look at. But it worked a lot better than I had expected. I could work Europeans easily on 20 and 40 meters with good signal reports. I worked up and down the East Coast of North America on 160 meters with good reports. That antenna's performance pleasantly surprised me.

HILTON HEAD ISLAND, SOUTH CAROLINA

On a mid-winter's vacation to Hilton Head, Maggie and I rented a cottage on a lagoon surrounded by large, old trees covered with Spanish moss. Sometimes you don't have enough trees (see New Brunswick, above) and sometimes you have too many....

The ham shack was in a second-floor loft area. I strung two wires out of windows on opposite sides of the room and ran them through the trees. One wire was about 75 feet, and the other, about 50 feet. I ran Wire 3, about 25 feet long, inside the house. The center of the antenna was about 6 feet above floor level on the second floor. That antenna was about as haywire as you can get.

Once again, the antenna loaded well on all bands. The higher-frequency bands were closed most of the time, so I didn't make many contacts on 20 meters and above. Conditions were good on 40 meters for rag-chewing by day and DXing by night, and I got out very well on that band. I made a few short-haul DX contacts with Caribbean hams on 80 meters and a couple of European contacts on 160 meters. Once again, the simplistic concept of running out three random-length pieces of wire served me well.

AVIEMORE, SCOTLAND

Maggie and I were spending a week in a timeshare condo in Scotland, the country her mother was born in. I took my ham station in a carry-on suitcase so I could get on the air a little from our condo in Aviemore. The condo had a deep back yard, and I was able to run a wire about 90 feet long from the second-floor ham shack to a tree out back. I ran two other wires about 25 and 40 feet long to other smaller trees that were nearer the building. The antenna loaded well on every band I tried, and contacts were easy to make — with good signal reports received.

SINT MAARTEN

Several years ago, Maggie and I found a bargain deal and bought a timeshare week in a condo in Philipsburg, Sint Maarten, right on the beach at Half Moon Bay. On our first visit to that timeshare, my friend of over 50 years,

Figure 19 — W1AB's timeshare condo is the fifth-floor corner unit shown in this photo of the Sint Maarten Sea Palace, on Half-Moon Bay in Philipsburg, Sint Maarten. (W3TMZ photo.)

PJ7/W1AB

Sint Maarten Sea Palace
Philipsburg, Sint Maarten

Figure 20 — The PJ7/
W1AB QSL card, with
the antenna sketched in.

Mike Jaquish, W4WJH, and his wife Merna joined us. I took my carry-on ham station with me for some HF fun.

Our condo is a corner unit on the fifth floor of the Sint Maarten Sea Palace. The balcony railing is about 40 feet about ground level.

One evening after dark, Mike and I put up what turned out to be a bodacious antenna. It's often best to install low-profile HF antennas after dark.... Mike went down to the beach, and I stayed on the condo's balcony. We stayed in touch with 2-meter handheld transceivers. I dropped the weighted end of the small-gauge wire to Mike, and he started strolling casually down the beach in a westerly direction, parallel to the beachfront. About 140 feet from the balcony, Mike found a convenient short pole, to which he attached the end of the wire. I tied off the other end of the wire to the wooden balcony railing. That became our Wire 1, oriented in an east-west direction.

Then Mike came back to a point just below the balcony, and I dropped the weighted end of another piece of wire into a palm tree adjacent to the balcony, while he watched to see that it "disappeared" in the foliage. That was Wire 2, and it came to rest vertically in the tree at a length of about 35 feet.

Wire 3 was an L-shaped wire, about 25 feet long, run from the side balcony toward the beach side of the balcony, then around the beachfront balcony. That version of the Killer Antenna turned out to be the best I've had to date in my portable operations. It was long, and the center was high. Radiation was mainly in the north-south directions.

Mike and I could work up and down the East Coast of North America on 160-meter CW (with occasional forays into the SSB mode). Two stations in Europe heard us on 160-meter CW, but we couldn't make any weak-signal contacts on that band because of the high manmade noise level. Dang!

We could work European and South American DX easily, with good signal reports, on 40 meters and the higher bands. We worked a few Asian stations, although conditions weren't very good in that direction.

I ran schedules with W3TMZ (in Florida) several times, and we had virtual pipeline conditions in the late evening on 20-meter SSB — big signals and no interference. I could get used to that!

CONCLUSIONS ABOUT THE KILLER ANTENNA

As described in the last few pages, I've used several variations on the original Killer Antenna in my portable operation here and there. In all cases — with widely varying antenna arrangements — the antennas have loaded well on all bands and I worked DX easily with good signal reports. You could consider using the Killer Antenna approach for your low-profile situation.

Appendix G of this book is a reprint of the *QST* article that described the original Killer Antenna.

9

Making the Wire Antenna Nearly Invisible

There are two main approaches to making low-profile HF antennas nearly invisible: (1) Use small-gauge wire that becomes virtually invisible at even moderate distances. (2) Conceal the antenna wire by blending it into existing structures.

Conventional ham wisdom has always taught us to use large-gauge wire for our HF antennas — the larger, the better. But what that conventional wisdom overlooks is that small-gauge wire works, for all practical purposes, as well as does the large-gauge. And it's much easier to hide.

Many years ago, I came across two spools of wire at a hamfest flea market, each of which with miles — yes, *miles* — of small-gauge insulated wire on the spool. They were being offered for sale at a ridiculously low price, because it's not the type of thing that hams would normally have any use for. The wire was seven-strand insulated wire, with a total wire gauge of 26. The wire on one spool had yellow insulation, and the other had brown insulation. I bought both spools.

Oh, yes — you remind me that conventional ham wisdom says you should not use insulated wire for antennas. This is another case where the difference between insulated and bare wire is inconsequential.

My small-gauge wire with brown insulation vanishes from sight quite well when you get 15 or 20 feet away from it. But if you really want to camouflage small-gauge wire, here's a

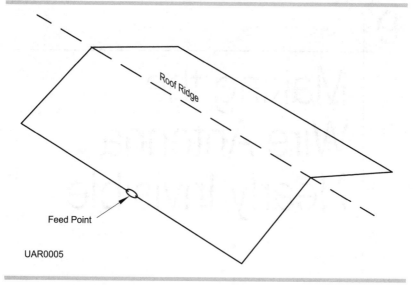

Figure 21 — It's easy to install a loop of wire around the underside of your home's eaves, and feed it at the middle of one side of the loop.

suggestion: Get some small-gauge wire (#26 or so, stranded) with black insulation. Buy a few small cans of paint — dark blue, brown, and dark green (or other appropriate camouflage colors of your choice). Wind about 150 feet of the antenna wire into a wrap of about 10 inches diameter. Now dunk portions of the wire into the various cans of paint, so the insulation will end up painted in various colors along its length. After the paint has dried, you can make an antenna out of the wire that will be virtually invisible at a distance of 15 feet or so.

A word of caution to the readers who are married hams: Do *not* do the camouflage painting on your spouse's kitchen counter or table.

If you are afraid the antenna won't go unnoticed as it crosses your yard about 20 feet above the ground, you can wrap the antenna around the outside of your house and tuck it behind or underneath parts of the structure. Most single-family homes today have very little metal in their structure (with the exception of homes with aluminum siding). You can run a loop of wire completely around the house, tucked underneath the eaves. A large-diameter loop like that can be fed either at a corner of the house, or halfway along one side of the house.

Also, see the article reproduced as Appendix A in this book for NØADL's approach to a hidden wire antenna. For ideas on installing antennas in your attic, see the article by W4KVS that is reproduced as Appendix B in this book.

You can often run wires vertically up the side of a tree and feed it at the bottom, working the vertical element against either an earth ground at the base of the tree trunk or against a counterpoise of wires buried beneath the ground at a shallow depth. If you use wire for the vertical elements that has insulation approximately the color of the tree bark, the antenna virtually disappears. See Appendix I for the description of a good multiband tree antenna that WB3BJF constructed.

By the way, if you want to be real cool, don't pronounce the word "tree" as normal people do. Convert that name into militaryesque nomenclature — call it a "tee-are-double-e."

Let me again suggest that you learn a lot about antennas, so you can look at your own situation and start thinking about what kind of wire you can put up to get on the air on HF.

10

Disguised Antennas

I found one ham who was having a really good time with his low profile and disguised antennas — Gene Preston, K5GP, of Austin, Texas. Gene and I swapped a couple of e-mails, and one of them was so informative, interesting, and fun to read that I'd like to share it with you, just as Gene wrote it. Gene's description and a photo of his low profile and very effective antenna field are in the sidebar beginning on the next page.

Don't you just *love* the way Gene thinks, and the nice low-profile antenna system he put up? Follow Gene's suggestions and get to work on your own low-profile antenna system!

Bill Glenn, AA4BQ, is another ham who uses a birdhouse mast to hide his antenna. The accompanying photo appeared in "Up Front in *QST*" in the July 2001 issue of the magazine.

Albert Parker, N4AQ, also disguised his antenna quite well. He told about his concealed antenna in the May 1993 issue of *QST*. Albert placed a Hustler 4-BTV multiband vertical inside a piece of PVC tubing to conceal the antenna, and then used the PVC tubing as a flagpole. See Appendix H to read the reprint of his article.

As you see, you can put your antennas in plain view....

Amateur Radio Station K5GP

The single R7000 Cushcraft vertical is behind the only good oak tree on my lot so as to be as invisible as possible to the neighborhood. I painted it a dark green color called "camouflage."

All the dipoles perform as NVIS (near vertical incidence skywave) antennas because they are so close to the ground (8 to 18 ft). The 40-meter phased dipoles design is posted at www.egpreston.com/40mbeam.txt. There is a picture of the pattern of the 40-meter phased antenna on the second page of the *PowerPoint* file posted at www.egpreston.com/RFItalk.ppt. The phased dipoles bring the angle of radiation down lower to overcome the bad feature of NVIS — that of not having a low angle of radiation.

But I still get the good feature of NVIS — low noise reception. The antenna is very quiet. When I run a kilowatt into it and can hear weak signals, I can work DX very easily on 40 meters. When I switch from the phased dipoles to one of the other antennas, the DX signal frequently slips into the background noise. The VSWR of the array is less than 2.0:1 on the whole 40-meter band. Interestingly, the antenna performs best if low to the ground, making it the perfect invisible backyard beam for a restricted neighborhood, if you have space to put it up.

I use 18-gauge Copperweld wire, purchased from The Wireman

Figure 22 — The K5GP low-profile antenna farm.

(via the Internet). I made the mistake of using little clear plastic insulators. They deteriorated in the sun and lost their tensile strength. They also arced over when wet, with me running a kilowatt on the antennas. I now use the grey fiberglass insulators that I scavenged from a multiband dipole antenna that I had previously used.

I use 80-lb-test fishing line for the insulated line from each dipole to the terminating trees. I have springs on the dipoles for 80, 40, and 20 meters to give a lot of stretch to allow for tree movement when the wind blows. The springs are not real stiff, with only a few pounds of tension and not stretched out much under no-wind conditions.

All the antennas are fed with a single run of LMR400 low-loss 50 Ω coaxial line. An MFJ four-antenna remote antenna switch selects the antenna in use. That switch sends the switching power through the coax, so that no extra wires are needed to power up the remote switch box.

Many of the neighbors know about the antennas and are not bothered by them because the antennas are nearly invisible. Here is what I did to get the antenna up and running in spite of the restrictions. Other low-profile hams can follow the same course of action.

1. Put up the hidden antennas and don't tell anyone. Asking for permission is a certain way to get denial before the antenna even gets off the ground.

2. If the antenna is noticed, try to work with the neighborhood to make it as invisible as possible. Ask to be granted an exception if pressed by the architectural committee of your neighborhood or condo association.

3. If possible, try to be a member of the architectural committee before your antenna becomes an issue. [Gene: that is downright sneaky! — Al] Currently I co-chair the architectural committee for our neighborhood and — believe me — there are a lot of more pressing issues than hidden antennas.

4. Look for other violations that have not been acted on, so you can strengthen your position if your antennas are challenged. Note any exceptions that have been made in the past to your CC&R. When the no-antennas clause was put in your contract, they were thinking of lattice towers and TV antennas on rooftops. They were not thinking of hidden ham antennas, which may be more acceptable to the neighbors than rooftop TV antennas would be.

My hidden antennas are now pretty well known in the neighborhood and no one has complained — not even my neighbor next door, who clearly can see the wires. That neighbor leaves his trashcan beside his house, which is a violation of the covenants... but I dare not complain — hi.

73 de K5GP, Gene

Figure 23 — AA4BQ's ingenious disguise for a Hustler 4BTV vertical makes it look like the support pole for a birdhouse. We only hope that this doesn't cause his signal to have chirp on it.

Figure 24 — Always be sure that your disguised antenna blends in well with its surroundings.

Fitting Conventional HF Wire Antennas Into Indoor Spaces

You can use the various types of wire antennas indoors that are described in ham radio's antenna literature (books and magazine articles). The main problem you will run into is sometimes not having enough straight-line clear space for a given antenna either in your living space or in your attic. The easy way to deal with those space limitations is to route the antenna so that it is no longer all in one straight line.

FORGET THE STRAIGHT LINE

Let's use the common half-wave dipole, center-fed with coaxial cable as an example. Look at the accompanying figure. (1) You can install the two halves of the dipole as an L-shaped antenna. (2) You can leave the center part of the antenna in a straight line and bend the ends of the antenna at whatever angle the available space dictates. (3) If you are installing the indoor dipole in a room in your living space, you can run the center of the dipole diagonally across the ceiling of a room, and then run the ends of it vertically down the corners of the room. (That scheme requires your spouse's approval, of course.)

There is no way to discuss all the variations you can make when running your dipole in something other than the customary straight line. Just keep in mind that you can usually get away with running it almost any way and it will still work.

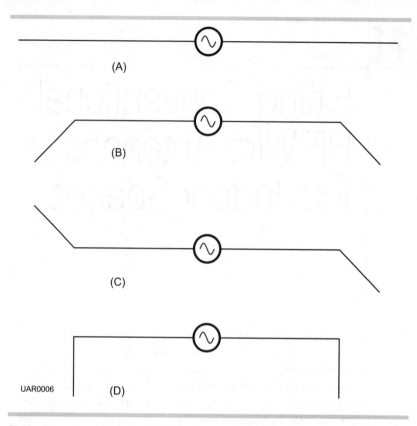

Figure 25 — Hams typically think of a dipole as being put up in a straight line (A), but it can be bent various ways (B, C, and D) to fit it into available spaces, such as attics.

WINDING UP PART OF THE DIPOLE INTO ONE OR A PAIR OF LOADING COILS

If necessary, you can wind part of the length of the dipole into a coil. If you do that, you can use a large-diameter (preferably at least 2 inches in diameter) round tube made of a nonconducting material as a coil form. You can either wind part of the length of one half of the antenna into a series loading coil, or you can make the antenna symmetrical by winding the same portion of each half of the antenna into two loading coils of the same size at the same distance from the center feed point of the antenna.

If you wind part of the antenna into a loading coil (either a single coil or a pair), about 20% of the length of the antenna from the center connection to the loading coil should remain straight, before reaching the loading coil(s). The reason for that is because the center of the dipole is the high-current portion of the antenna, and therefore is doing the bulk of the radiating of the signal. The 20% is my rule of thumb; if necessary, you could have less than that length remain straight.

If you are using uninsulated wire for your antenna, you will need to keep adjacent turns from touching each other. One easy way to accomplish that spacing would be to wind both the antenna wire and small rope (about the same diameter as the wire) side by side as you wind the coil.

Making your own loading inductances by winding up part of the antenna into coils will almost certainly change the resonant frequency of the antenna. If your VSWR meter shows that to be the case, you can splice additional length of wire to each end of the antenna to lower the resonant frequency, or cut off equal lengths of wire from each end of the antenna to raise the resonant frequency.

ANOTHER PLUG FOR LADDER LINE

If you use 450 Ω ladder line or 300 Ω twin lead instead of coaxial cable to feed a contorted dipole at its center (via an antenna tuner with balanced output), the situation becomes simpler. If you don't have enough space for the full length of the dipole, then you can cut its length shorter — and even make it nonsymmetrical, with one end longer than the other (so it becomes an off-center-fed dipole). The antenna tuner will be able, in almost all cases, to

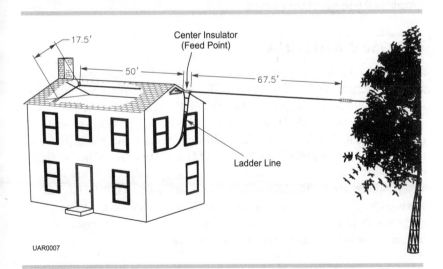

Figure 26 — W1AB used this all-band HF dipole from his third-floor apartment in New Britain, Connecticut. Kirk, NTØZ, came up with the catchy name "HIHO Attic Special" (half indoors, half outdoors) for it. If you can run the ladder line inside the building, all that will be exposed to view will be the HO part of the dipole. Using small-gauge wire for the HO half will make the antenna virtually invisible. This is a very effective low-profile antenna.

match the feed line to the transceiver.

An antenna that I used for a few years from a third-floor apartment is another possibility. See the accompanying figure.

That antenna was a 130-foot dipole, center fed with ladder line. I placed half of the dipole in the attic, one floor above my shack. The depth of the building (front to back) was only about 50 feet, so I bent the indoor half of the dipole as shown in the figure. The ladder line was entirely indoors — it couldn't be seen by passers-by. (I didn't have any CC&R restrictions, but the antenna just happened to be a low-profile antenna.)

The half of the antenna that was outdoors was made of the 26-gauge stranded, insulated wire that I've described elsewhere in this book. That wire was about 25 feet above the ground — people standing on the ground had trouble seeing the 26-gauge wire at that distance.

I fed the dipole using an antenna tuner with balanced output. It was a very effective all-band HF antenna that was very difficult to see, even when you were looking for it.

WIRE ANTENNAS OTHER THAN CENTER-FED DIPOLES

There are several other all-band antennas that shrink the length of an antenna for 80/75 meters. Three examples are the G5RV antenna, the Windom antenna (fed with 300 Ω twin lead), and the "trap dipole" that uses LC tuned circuits in the length of the antenna.

THE G5RV ANTENNA

You can find descriptions and discussions of the G5RV antenna in the antenna literature. The original G5RV design was a center-fed dipole that used a section of matching line made of 34.8 feet of 450 Ω ladder line between the 102-foot dipole and the 50 Ω coaxial feed line. Studies have shown there is no length of balanced line of *any* characteristic impedance that would perform the desired impedance transformation to 50 Ω on all HF bands. G5RV himself recommends using an antenna tuner to feed the 450 Ω ladder-line feed line.

However, like the bumblebee, many hams still use the G5RV with coax and a ladder-line matching section, and they swear it is a terrific antenna. I suggest to them that they also erect a half-wave dipole and get comparison reports on the two antennas to see which works better.

THE FAMOUS WINDOM ANTENNA

The Windom antenna became popular in the 1930s (before coax became available at low cost), and was in progressively decreasing use by hams into the late 1950s. It's a 130-foot flattop fed with a single wire that's attached to

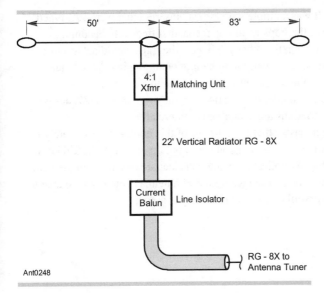

Figure 27 — The Carolina Windom antenna works on all HF bands, 80 through 10 meters. See **www.radioworks. com/ccwcover. html** *for a good description of the Carolina Windom and its variations.*

the antenna 47 feet from one end. That makes it a good candidate for a low-profile antenna.

Then a variation was made to the original Windom antenna, feeding it (at a slightly different point along the length of the flattop) with 300 Ω twin lead.

Later still, three hams from the Low Country of North Carolina and Virginia came up with the idea of the "Carolina Windom," which you can build yourself or buy from ham suppliers. The accompanying figure shows the Carolina Windom concept.

THE TRAP DIPOLE ANTENNA

The trap dipole antenna, developed by Chester Buchanan, W3DZZ, was first described in the article, "The Multimatch Antenna System," in the March

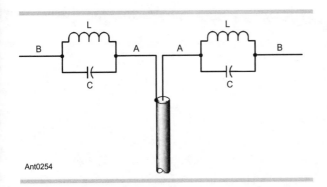

Figure 28 — The basic configuration of the W3DZZ trap dipole. Later variations of the trap dipole added a second pair of traps to the antenna.

1955 issue of *QST*. It was a dipole with "traps" inserted that were L/C tuned circuits. On one band, the traps would be resonant, effectively becoming insulators on that band. On the other HF bands, the traps would act as either inductors or capacitors, electrically lengthening or shortening the antenna (respectively) to bring the dipole into resonance on those bands.

There have been a number of variations on the original W3DZZ design, and many of those variations are available commercially.

Let me again emphasize that the purpose of this book is to stimulate your thinking about what kind of antenna you could put up in your CC&R'd home. You might not be able to duplicate the antennas that are described in the ham antenna literature, but you can see what kind of antennas you might be able to use and escape the surveillance of the homeowner associations.

12

Antenna Tuners

Over the years, there have been hundreds of thousands of words written in the ham literature about antenna tuners. Get ready; here come a few more.

Antenna tuners (or, as some prefer to call them, *antenna system tuning units*) are passive devices that transform the impedance of the antenna system to the transceiver's design load impedance, usually 50 Ω.

One important consideration about antenna tuners is that they should not be *lossy*. It is possible, when using an antenna tuner, to arrive at settings of the tuner controls that give a low SWR reading but where much of the transmitter's power is being dissipated in the antenna tuner itself rather than being fed to the antenna system. If you're using an antenna tuner and it becomes warm to the touch, its efficiency is not good.

It's also possible to develop high RF voltages inside the tuner, even with power outputs of 100 W or less. In the worst case, you can cause arcing inside the tuner. For that reason, it's always good to buy or build an antenna tuner with power ratings at least three times the power you expect to run through the tuner.

In this chapter, we'll look at manual antenna tuners. We won't look at *autotuners* that are located directly at the antenna and connected to the transmitter by a run of coaxial cable.

Until the past few years, the typical antenna tuner has been an *unbalanced* device. A toroidal balun coil would be placed at the antenna end of the tuning circuitry to provide balanced

Figure 29 — A look at the front panel, rear panel, and interior of the Palstar AT1500CV antenna tuner, rated for operation at 1500 W on 160 through 10 meters.

output for feed lines such as ladder line or twin lead. At some frequencies, the balun could cause additional loss, evidenced by its becoming warm to the touch during operation (feel it with the transmitter on standby, not when it's transmitting).

If you have an unbalanced antenna tuner that does *not* have a balun built into it for feeding a balanced feed line, you can add an external balun at the antenna tuner's output terminal. The accompanying photo shows a balun I wound to use with a Dentron antenna tuner that I sometimes use. I connect

Figure 30 — This homebrew toroidal balun connects to the output of a single-ended antenna tuner to feed a ladder line feed line. The ladder line clips onto (1) the end of the black wire and (2) the white wire where the insulation has been removed, directly behind the PL-259 connector.

the PL-259 connector to the antenna tuner's output and use alligator clips on the end of the feed line to clip onto (1) the end of the black-insulated wire and (2) the spot on the white-insulated wire where the insulation has been cut away, directly behind the PL-259. This balun isn't a thing of great beauty, but it works well.

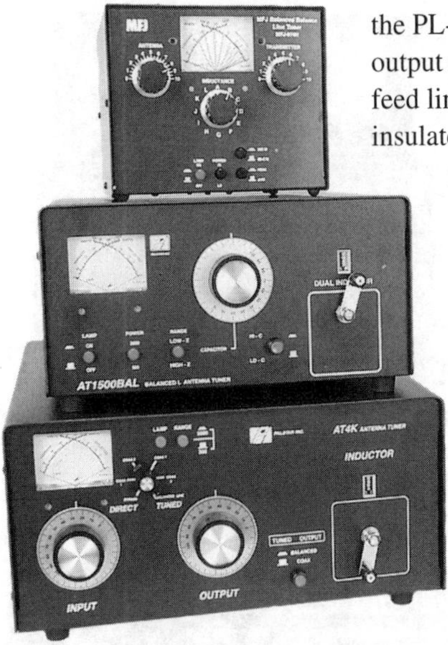

During the past few years, a new generation of antenna tuners has appeared. Those tuners have a toroidal balun at the input of the tuner, followed by a tuner circuit that provides balanced output. See the "Product Review" in the September 2004 issue of *QST* for a look at some of the current balanced antenna tuners, as well as a discussion of balanced antenna tuners.

Figure 31 — A stack of three balanced antenna tuners (top to bottom): MFJ-974H, Palstar AT1500BAL, and Palstar AT4K.

Hundreds of antenna tuners have been described in the ham literature. You can build your own antenna tuner from the information provided in those articles. Building an antenna tuner is interesting work and you usually learn things by doing the project, so I recommend that you consider that as a possibility.

Two interesting antenna tuner circuits have appeared fairly recently in articles in *QST*. I recommend that you consider them if you plan to build your own tuner: "A Compact 100-W Z-Match Antenna Tuner," by Phil Salas, AD5X, *QST*, January 2003. "The EZ-Tuner," by James Garland, W8ZR, is described in *QST*, January 2002. In addition to these two, you can search for antenna-tuner articles on the *QST* Web site and find plenty of reading to stimulate your thoughts about your own low-profile station and its antenna requirements.

If you prefer to buy a commercial tuner, there are many ready-made units on the market. The "Product Review" column in

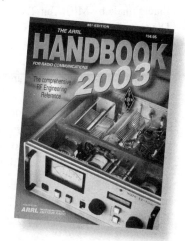

*Figure 32 — A view of W8ZR's EZ-Tuner graced the cover of the 80th Edition (1993) of **The ARRL Handbook**, as well as the cover of the May 2002 issue of QST.*

the February 2003 issue of *QST*, "*QST* Reviews Five High-Power Antenna Tuners" is a worthwhile read, with good technical discussion of antenna tuners in general. By the way, if you plan to buy a commercial antenna tuner, you should consider buying one with higher power ratings than you need for your current station. Later on, you might want to run higher power (especially if you move out of your CC&R'd home), and you would already have an antenna tuner suitable for the task.

During most of the many years that I've been a ham, I've used feed lines other than coaxial cable, together with an antenna tuner to keep everything happy. As a matter of fact, the second article of mine that was published in *QST* was an antenna tuner I had designed and built ("A Five-Way Antenna Coupler," *QST*, November 1958). Many of the antenna tuners I've used through the years have been ones that I homebrewed, but I've also used tuners made by MFJ, Ten-Tec, Dentron, E. F. Johnson, and others. All those tuners have served me well, I've had good results with them, and I seldom had any problems with them.

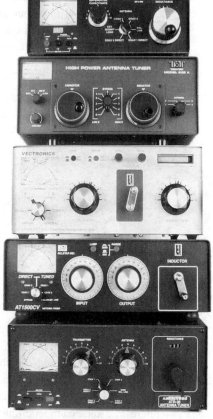

Figure 33 — A stack of high-power antenna tuners (top to bottom): MJF-986, Ten-Tec 238A, Vectronics HFT-1500, Palstar AT1500CV, and Ameritron ATR-30. Even though you might run low power today, you can use a high-power tuner with your current rig...and be ready for the day when you get a linear and go to higher power.

I learned one trick early on, when using an antenna tuner with only one antenna to operate on several ham bands. I would make up a simple table of control settings to enable me to quickly return to a previously used frequency. When switching bands, I would first preset each control to the value listed in

Table 1
Cheat sheet for my Ten-Tec 238 antenna tuner and W1AB Killer Antenna

160	2.0	High-Z 2	135		20	3.5	Low-Z 1	13
80	8.5	High-Z 2	156		17	2.5	Low-Z 1	52
75	2.0	High-Z 2	156		15	1.0	Low-Z 1	7
40	8.5	High-Z 1	57		12	1.0	Low-Z 1	0
30	9.5	Low-Z 1	20		10	0.5	Low-Z 1	8

Figure 34 — My first antenna tuner, and the tuner in place in my 1958 station (Collins 75A2 receiver, Eldico EE-2 keyer, Viking Ranger, antenna tuner).

Table 1 for that band. After that was done, I would close the key and quickly touch up each control for minimum reflected power. Thus I could have the antenna system tuned while radiating an on-the-air tuning signal of only about 10 seconds duration. That's a good way to keep the bands clear of the tuner-upper problem.

For further reading on the subject of antenna tuners, which I highly recommend, check out the articles listed at **www.arrl.org/tis/info/tuner.html**.

13

Variations on the Screwdriver Antenna

Let's consider a generic *Marconi antenna*. The Marconi antenna in its most basic form is a quarter-wavelength antenna that is operated against a counterpoise (or earth ground) and is fed at its base with 50 Ω coaxial cable. The counterpoise (or earth ground) provides an RF mirror effect so that the antenna then acts as if it is a half-wave dipole. To visualize this, place a small mirror flat on a table, and balance a pencil (which we will call a quarter-wavelength antenna) vertically in the center of the mirror (or hold it, if you can't get it to stand up). Now look at the pencil and mirror image from the side at about a 45° angle. What do you see? Dang! You see a *half*-wavelength antenna made up of a quarter-wavelength antenna and a quarter-wavelength reflection.

Now getting back to the real world…. In cases where a vertical antenna must be shortened to fit within a height limitation (such as a mobile antenna), an inductor can be inserted in the vertical antenna to make the antenna act as if it is much longer than the actual physical length. As was discussed in Chapter 6, the positioning of that loading coil along the length of the vertical element will affect the radiation efficiency of the antenna.

Looking at the screwdriver class of mobile antennas, we see that the main advantage of the screwdriver antenna when compared with either the Hamstick type of antenna or a Hustler mobile antenna is that the screwdriver antenna can be tuned to any frequency in any HF ham band from the driver's seat, while the vehicle is in motion. That is possible because of the remotely tuned and continuously adjustable loading coil. That is a big advantage, when you consider that the usable bandwidth of a Hamstick or Hustler mobile antenna might be as little as 25 kHz on 80 meters, and perhaps 40 kHz on 40 meters. A narrow usable bandwidth like that is okay if you have a favorite operating frequency and stick pretty close to it. But if you're like most hams, you will want to tune up and down the band, to see what's going on, on various frequencies. The Screwdriver antenna to the rescue!

When you put a HF mobile antenna in your attic for low-profile hamming from your home, the same advantage applies. You certainly don't want to be climbing up to your attic and adjusting the stinger (short top whip) on a Hamstick or Hustler when you move up and down the band.

You can put a screwdriver mobile antenna in your attic and use the remote control to vary the inductance of the loading coil of the antenna. You can quickly and easily move up and down a given HF ham band, and you can change from one band to another from the comfort of your shack's easy chair.

The screwdriver antenna must have a *counterpoise* to work against. A simple counterpoise can be made from several wires in your attic. Connect the counterpoise wires to the shield (ground) side of the coaxial cable feed line and run them along the attic floor. A counterpoise system can be made from one wire (or more) cut to a quarter-wavelength for each ham band you want to use. However, it's also possible to make a counterpoise from wires of various random lengths (between 8 and 66 feet long for operation on all the HF bands). In either case, the counterpoise wires should be run out from the base of the screwdriver antenna.

If you don't have enough space to run the long counterpoise wires straight out from the antenna base, you can either make a series of S-turns with them to take up some length, or you can run them straight out as far as possible and then bend them as necessary to fit within the attic space. Just remember to secure the counterpoise wires in your attic, so you don't trip over them later. One easy way is to run a length of duct tape along each wire and tape the wires to the attic floor.

Now let's consider a minor but significant modification to the screwdriver mobile antenna. Remember that the bottom of a vertical antenna (one that is an electrical quarter-wave in length, as is a screwdriver antenna with a loading coil) is where the maximum RF current is flowing, and where the maximum antenna radiation is taking place. The RF current (and radiation) becomes less as you move toward the far end of the antenna.

Because of that, it is advantageous to place the loading coil fairly near the top of the antenna. Conversely, you can extend the *bottom* of the attic-mounted screwdriver antenna, thereby displacing the loading coil upward along the antenna's length.

Most (if not all) of the commercial screwdriver antennas are terminated at the base with a standard ⅜-24 stud. It would be possible to fabricate a bottom antenna extension section to extend the length of the antenna. But the easiest way to extend the bottom length of the screwdriver antenna is to buy some masts with a ⅜-24 female thread on the top and a ⅜-24 stud on the bottom.

Aha! How about the Hustler MO-3, a 54-inch solid mast with the desired ⅜-24 connectors at the top and bottom? That's exactly what you need. Or, if you have restricted vertical clearance in your attic, you could use a Hustler MO-2 fold-over mast — the bottom 27 inches of your extended screwdriver antenna would be vertical, with the rest of the antenna folded over into a nominally horizontal position.

Shoot, you could even put *two* Hustler masts below the screwdriver antenna — an MO-3 at the bottom, then an MO-2 above that. To do that, you would need 81 inches of vertical clearance in your attic. If you don't have that much space, then you could put the MO-2 at the bottom and the MO-3 above that. Keep the brain in gear and look at all the possibilities that are available to you!

Obviously, the higher you can place the loading coil along the antenna's overall length, the better its radiation efficiency will be. But wait a minute, you might ask, what will happen with the tuning of the screwdriver antenna if you add length at the bottom of the antenna? With the added antenna length, less inductance will be required to tune to a given frequency, as compared to the tuning of the original screwdriver antenna. That's no big deal — all you're doing is varying the loading coil's inductance until the antenna resonates on frequency. It will resonate at the same frequency with a slightly different inductance setting — that's all.

If your attic space has a really low overhead, your modified screwdriver antenna can be mounted in the horizontal plane, or at whatever convenient angle is available. No matter what position it's placed in, it will radiate and you will be surprised at how many good contacts you can make with such a small and simple antenna.

Phil Salas, AD5X, came up with a nice little gadget to use with a Little Tarheel screwdriver antenna. The gadget is a turns counter that shows when you have reached either end of travel of the adjustable loading coil. The accompanying photo shows the compact unit. For details on building and operating the switch, as well as more photos of the switch, see Phil's article, "Tuning Switch with Limit Indication for Screwdriver Antennas," in the Dec 2005 issue of *QST*.

Compact Loops for HF

You can build your own compact loops quite easily. Rod Newkirk, W9BRD/VA3ZBB, used homebrew compact loops for many years from his home in Chicago. He learned how to build the loops by reading the literature about them and then by a lot of cut-and-try experimentation. Rod wrote the following description of his compact loop techniques, which was first published in 1993 in the ARRL publication *Low Profile Amateur Radio*, by Jim Kearman, KR1S:

> The circumference L (see the accompanying figure) of the compact loop should be somewhat shorter than a quarter wavelength, about 25 feet for 7 MHz. C1 (300 pF) and C2 (300 pF) should be breadboard mounted and within easy reach from your operating position, so you can retune the loop as you change frequency across its frequency band. The antenna system bandwidth is only about 50 kHz. Midget receiver-type capacitors are OK for powers up to about 150 W.
>
> Because the feed point is unbalanced, some RF may appear at the ground point. A counterpoise consisting of a quarter-wavelength of insulated wire should be attached to the ground point. It can be run along the baseboard or under a rug, to keep it out of sight and to avoid a trip hazard. At power levels of 100 W or more, the voltage at the far end of the counterpoise may be high enough to generate a corona arc. Check for arcing at the highest power you intend to

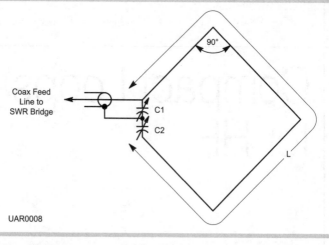

Figure 35 — W9BRD's simple matching system for a compact loop antenna. C1 is 300 pF and C2 is 100 pF. The loop diameter should be slightly less than one-quarter wavelength (eg, 25 feet for 40 meters). Although this diagram shows the loop to be diamond-shaped, you can alter the shape to fit your available space.

use before hiding the counterpoise.

The 25-foot circumference is good for 40 meters, but is too large for the higher bands. However, the 25-foot loop dimension can be used for 30 and 20 meters by inserting a third capacitor (300 pF) at the end opposite the feed point to capacitively shorten its effective length. To use the loop on other bands, you can install other quarter-wavelength counterpoise wires (one for each of the additional bands) in parallel with the first counterpoise.

Similarly, the resonant frequency of any given loop can be raised slightly by inserting a third capacitor (of about 100 pF) at the corner of the loop opposite the feed point. For frequencies above 14.35 MHz, the 25-foot loop is too large for the C1-C2 bridge to do its job, even with C3 at minimum capacitance. You can build a smaller loop, of course. But another possibility is to remove the capacitors, open the loop at the far end, and tune it with a balanced antenna tuner.

The loop becomes progressively more directional as the operating frequency is increased. Having the loop open at the higher frequencies with the top and bottom fed out of phase sends more RF straight up and down than in the desired horizontal directions.

You can use a slightly smaller loop (22 feet in circumference) to function in a balanced mode from 18 to 30 MHz. Equal capacitances at the top and bottom tune the loop to the desired frequency range. You can substitute a gamma-matched quarter-wave stub made from twin lead for the capacitive bridge coupling. This antenna can be mounted in the vertical or horizontal plane. The two resonating capacitors are made from short pieces of 300-Ω twin lead. You will have to trim the lengths of the twin lead for resonance, with the strips as nearly equal in length as possible (a simple but time-consuming and tedious task).

Without the twin-lead tuning capacitors attached, tune a nearby receiver to the frequency you want to tune the loop to. Bring a dip meter near one of the closed corners of the loop and adjust the frequency of the dip meter so you can hear the dip-meter signal in the receiver. Then attach the pair of twin-lead capacitors and alternately clip each piece of twin lead (keeping them to the same length) until the dip meter indicates resonance. Their lengths are fairly critical.

Typical lengths for these twin-lead capacitors are 8 inches at 28 MHz, 13 inches at 24.9 MHz, 23 inches at 21 MHz, and 38 inches at 18.07 MHz. Each band also requires a matching stub of twin lead, which should be slightly shorter than a quarter-wavelength, and a separate preset gamma capacitor for each band.

At 21 MHz, for example, with the proper twin-lead capacitors in place, connect a 12-foot or longer length of twin lead across one of the strips. It's convenient to connect the strip to the top of the loop, with the strip coming straight down. Stick a straight pin into each side of the twin lead about 11 feet from the feed point and short the pins with a short jumper. Check system resonance as before, with the dip meter held near one closed corner of the loop. Move the pins up and down the stub until the system resonates within the desired frequency range. Remove the pins, cut the twin lead at that point, and make a permanent short.

The gamma match is easier to make. Clip the coaxial cable's outer braid to the shorted end of the twin-lead stub. Connect a 100 pF variable capacitor to the center conductor of the coaxial cable. Connect the other side of the capacitor to a pin stuck in the twin lead about 12 inches above the shorted end. Adjust the capacitor for minimum reflected power, as indicated on a VSWR bridge. If you can't tune the capacitor for zero reflected power, move the pin up or down the twin lead 2 or 3 inches at a time until you find the point where you can obtain zero reflected power. The correct matching point will likely be between 8 and 10 inches at 28 MHz and between 16 and 18 inches at 14 MHz.

Varying the spacing along the gamma wire (from the capacitor to the pin) will provide fine tuning. That spacing should nominally be about 1 inch. The position of the gamma wire must be kept stable — movement of that wire will change the tuning of the gamma match.

Once the correct gamma-match setting in determined, you can hold the gamma wire in place with tape. Then remove the pin, strip the insulation from the twin lead at that spot, and solder the gamma wire in place. That completes the tuning process.

The balanced 22-foot-circumference antenna works well on the bands down to 7 MHz, but at that frequency the twin-lead stubs are so large that they become unwieldy. On 14 MHz you can resonate the loop with a pair of identical midget variable capacitors (40 pF each) and couple the antenna via a twin-lead stub. At 10.1 and 7 MHz, you can use two other pairs of variable capacitors (60 and 100 pF, respectively) in the balanced capacitor bridge arrangement described earlier.

To maintain loop balance, it's best to make an RF choke just below the bridge by winding 20 turns of the coaxial feed line tightly on a ½-inch ferrite rod.

Although so large as to be almost impractical, you could make a loop for 3.5 MHz. The loop circumference would have to be about 50 feet, with a counterpoise about 60 feet long. A loop of that size can be made by running the loop wire around the circumference of the wall in an average-sized room. The capacitances would have to be about 250 pF each.

Those guidelines of Rod's will help the experimenter build compact loops for the various HF ham bands, 7 MHz and higher. Or down to 3.5 MHz if you feel adventurous!

Additional information on building compact loops can be found in the article, "Small, High-Efficiency Loop Antennas," by Ted Hart, W5QJR, in the June 1986 issue of *QST*, and in the section titled "Small Loop Antennas" in the "Loop Antennas" chapter of *The ARRL Antenna Book*.

15

HF Operating Tips

As a long-time participant in and observer of Amateur Radio, I offer my opinion that the operating skills of the average HF ham today aren't as good as those of the hams of 50 years ago. The process by which hams today become licensed and get on the air often skips over the mentoring process that occurred with hams who entered ham radio those many years ago. That's too bad, because the mentoring process is where the newbie hams of yesteryear learned a lot of good effective operating techniques and thus became the good operators of the present. Not having that mentoring process is akin to being thrown in the deep end of the swimming pool to learn to swim. You flail around a lot, and you likely don't come near drowning, but it might be a long, tedious time before you become a swimmer with good technique.

One positive thing you can do is to identify a nearby ham with excellent operating skills and ask if you can come watch him or her operate their stations, either in casual operating or in contests. If they agree to host you, watch and listen to what they do as they operate. If they do something you don't quite understand, ask about it. That can help you pick up some good operating tips.

Then if the ham you have asked to be your mentor is agreeable, have her or him come and watch you operate and give you a critique. At this point, you must set aside any feelings of embarrassment, but instead be open and receptive to suggestions and tips your mentor offers.

Also, as you operate, or as you spend time listening to others hams on the air, identify the more successful operators and figure out what things make them successful. Note their effective techniques, and figure out how to use those techniques in your own operating.

CASUAL HAM RADIO

Let's first consider what I call *casual ham radio*. This concept includes operating, contesting, DXing, whatever — but doing it using a relaxed, casual approach. You operate without the expectation that you will get a reply to every call you make; without hoping to get a 59+ or 599 signal report on every contact; without running up a contest score that is in the Top Ten; without working every DX station you hear. You do it for the enjoyment of the activity. With more modest expectations, you can enjoy the thrill of your successes, without being disappointed by the occasional failure.

Casual ham radio is simply ham radio for fun and enjoyment — without getting into the highly competitive ranks of hamdom. There's a lot to be said for enjoying ham radio in this simplified manner.

Operating with restricted antennas almost automatically places you among the ranks of casual operators. *Almost*. The ham with restricted antennas has trouble competing with the hams who don't have CC&R limitations to deal with. However, over the years, many hams operating with simplistic or limited antennas have been very successful. One who comes immediately to my mind is Howie Bradley, W2QHH, now a Silent Key.

Howie was a Post Office mail carrier who did a lot of DXing and certificate-chasing in the 1950s and '60s. He used modest power and simple wire antennas from his home on a small residential lot in Hamilton, New York. Yet he worked almost all DXpeditions that showed up on the ham bands. He worked 352 countries (including 105 countries on 160 meters, back when 160 wasn't really considered to be a DX band), usually running 20 W to a Johnson Viking Ranger and never running more than 50 W. He qualified for virtually every operating certificate that was offered back in those days.

Howie supported communication with the Joint Arctic Weather Stations (an early effort to help weather forecasting — before the days of satellite weather imaging) from 1947 to 1955, providing official as well as personal message-handling service. Howie was a credit to Amateur Radio.

Howie did it all simply by making the most of good equipment and simple antennas, and by learning to be a consummately good operator. Howie, we miss you…but we hope some of the readers of this book will learn to operate as effectively as you did!

But let's leave the Top Ten contesters' lists, the Honor Roll, the superlative accomplishments in ham radio to those who don't have to deal with the limitations placed on hams by CC&R. Let's just get our low-profile ham radio station set up; our disguised, discrete, or invisible antennas installed; learn to be good operators (and learn all the good operating tricks) and enjoy the magic of ham radio operating!

For those hams who want to be as competitive as possible, it's necessary to learn a lot of the good operating tips. Here are a few, to stimulate your thinking about effective operating.

SOME OPERATING TIPS

Here's a first piece of good advice for the casual ham operator and the low-profile operator: Don't spend so much time calling CQ! Instead, listen a lot, find someone else calling CQ, and reply to his call. Your QSO rate (in terms on how many contacts you make during a given operating period) will likely be higher by doing that. Furthermore, by listening and identifying stations who are calling CQ, you might find a new state you would want to work, or a new country you haven't yet contacted, or some ham friend you've enjoyed talking with in earlier contacts.

Another tip: Be an interesting conversationalist. Keep a road atlas of the USA and Canada at hand (you can get the excellent *Rand-McNally Road Atlas* at Wal-Mart stores for $5). When you contact a ham in North America, look up his location in your road atlas. You might see a nearby tourist attraction, or lake, or river or something you could ask the other ham about in order to start a pleasant conversation about his locale, and thereby learn more about his part of the country. To be an interesting conversationalist, you need to open your ears and mind and encourage the other ham to tell you about where he lives, what he does for a living, where he has traveled on vacation or on a job assignment, etc. Just as in face-to-face conversation, you should be a good listener.

If you learn to be an interesting conversationalist, you will enjoy much more interesting contacts. Of course, when you get into conversations like that, the other ham will usually reciprocate and ask you similar questions, so you can brag a little about your town, your family, your job and your travels.

Like everything else in life, you can enjoy ham radio more by taking time to smell the roses. That's another advantage of casual hamming.

A follow-on advantage of being an interesting conversationalist on the air is this: The other ham will more likely remember you and your call sign, and he might make a greater effort to contact you at a later time for another interesting rag chew if he hears you on the air again.

Here's another operating tip that will help if you're trying to work a station through a pileup. Put yourself in the other operator's place. *Before you call him,* listen to his operating style carefully, to determine which other stations he's coming back to. Determine what kind of calling technique he's listening for — short or long (in pileups, the sought-after ham usually listens for short calls). Determine what frequency he's listening on. Often, the sought-after ham will *specify* what frequency he's listening on. But if he doesn't specify his listening frequency, you will need to listen carefully to determine where he's listening — whether on his transmitting frequency, on the same frequency for every contact, on a different frequency for each contact (and, if he's doing that, is he listening to progressively higher frequencies, progressively lower frequencies, or frequencies at random?), etc. If he specifies calling stations by call-sign area, do *not* call him when it isn't your turn. Give him a call when, where, and how we wants, and you stand a much better chance of being successful and working him.

Here's an example from my own experience of how to call when, where, and how the DX station is listening. One Sunday evening in the 1980s, I was driving northbound on I-270, heading home from Washington. I flipped on my Century 22 mobile CW rig and tuned it to 20 meters. The ARRL CW DX Test was in its closing hours, and I started working DX stations in the contest to pass the time. That late in the contest, there weren't many pile-ups on the DX stations, so I was usually working each station with only one call.

Then I came across HZ1AB, who was doing a great job of working through a humongous pile-up. It was unlikely that he would hear me in all that mess of RF, but — what the heck — I called him a few times anyway. Sure enough, he didn't hear me.

Then HZ1AB said, "QRX 10" and went on a short break. The frequency fell dead silent…absolutely and totally quiet. You could have heard a dit drop.

I glanced at my watch to check the time, and kept driving up I-270. The frequency remained totally quiet. You would have thought that particular spot in the otherwise-busy band was dead.

Again checking my watch, exactly nine and a half minutes after HZ1AB's "QRX 10" transmission, I sent "HZ1AB de K3KMO $\overline{AR}$" — on a totally quiet frequency. As soon as I finished that call, I heard "K3KMO de HZ1AB 599." I replied, "R 599 HZ1AB de K3KMO TU."

The frequency then became awash with RF as several hundred high-power, huge-antenna stations again starting calling HZ1AB. Meanwhile, I was still driving up I-270, but now I was literally laughing out loud at having put a rare one in the log so easily!

All I did was to put myself in the other operator's place, and then call him on the frequency, at the time, and in the manner, that he's listening for. Okay, I admit it was a rare situation — but you get the point. But let's consider the other side of that coin — where a ham is calling a DX station at the wrong time, on the wrong frequency, etc.

On far too many occasions, I've heard US stations calling a DX station over and over, making long calls and even transmitting at the same time (and often on the same frequency) as the DX station. It's obvious that the Frothing-At-the-Mouth caller (abbreviated here as FATmouth caller) hasn't even *heard* the DX station, has no idea where to listen to hear him, but is just wasting his time — and causing unnecessary interference and annoyance to the other, better operators on the frequency.

Do you think this kind of FATmouth calling will ever get a reply? *I don't think so!* If the DX station replied, would the FATmouth caller hear him? *Not a chance!*

In summary, learn to be a good operator. Read the ham literature to pick up good operating tips. Listen on the air to other successful operators, figure out what they are doing to be successful, and learn to use similar techniques. Better yet, visit nearby hams that you know to be good and successful operators, and watch and listen to them as they operate.

If you learn well from those good examples, you might surprise yourself with your own increased success rate.

16

QRP, or Not QRP; That is the Question

The ham who is forced to use low-profile antennas often is a ham who lives in an apartment or a condominium. Life in something other than a single-family, stand-alone house often raises further problems in ham operating.

For example, many consumer devices on the market today are quite susceptible to RF interference from HF ham rigs. Generally speaking, the lower the purchase price of electronic equipment, the more likely it is to be susceptible to interference from ham transmitters. And in what cost class are the electronic devices that most people buy nowadays? Yep, you got it!

Some of the typical devices that are RF-susceptible are stereo systems, electronic keyboard musical instruments, touch-controlled lamps, wireless doorbells and the inexpensive (read: *cheap*) wired telephones. Interference to TV sets isn't so common these days, for a variety of reasons. However, if your neighbors are picking up TV broadcast signals from distant TV stations (where the TV set's AGC is running the RF amplifiers at their highest gain), your HF signal might cause TVI problems, especially on the lower VHF channels (channels 2 through 5).

Therefore, it isn't the best idea for an apartment- or condo-living ham to run maximum legal power from his HF station. You can try running your rig at its usual 100-W output to see what happens, but you should first do some tests with a sympathetic neighbor (if you have one), to determine if you cause any RFI to any of his electronic equipment. From those tests, you can make an educated guess about whether you can run 100 W without the likelihood of RFI to your other neighbors.

Be alert to the possibility that, even after running tests with a given neighbor, you might still cause RFI to *other* neighbors.

If you do, your best course of action is to assume the responsibility for the interference and cure it from your end.

I know; I *know!* Your ham station is most likely clean, and the interference is probably generated within the poorly designed consumer electronic gadgets. Legally, it's not your fault. But your neighbor won't understand that — all he knows is that when *you* are on the air, *he* receives interference from your transmissions. It's just easier to do what you have to do at your end to eliminate the interference. Which gets us around to the matter of "QRP, or not QRP?"

QRP operation (5 W or less) has become progressively more popular in the HF ham community over the past 20 years, for a variety of reasons. One reason is that many hams enjoy taking small, lightweight portable stations with them on vacation or camping trips, especially backpacking trips.

Another reason is that a lot of hams enjoy the challenge of making contacts and the thrill of earning operating awards (such as WAS, WAC, and DXCC) using a transmitter whose output is about the same as the power consumed by the typical incandescent nightlight. Yet another reason — and sometimes the deciding factor — is the greatly reduced likelihood of the QRP rig causing interference to neighbors' consumer electronic devices.

Early manufactured or kit QRP transceivers were fairly simplistic, often using a direct-conversion receiver. Direct-conversion receivers have the advantage of simple circuitry, but they don't perform as well as conventional single- or double-conversion superhet receivers.

Current QRP transceivers have become electronically sophisticated, and many of the receivers in those rigs are as good as the receivers in the average 100 W ham transceiver. That is a significant advance, because the old saw, "You can't work 'em if you can't hear 'em," is as true today as it was in the early days of ham radio.

Running QRP rigs on HF, especially when using low-profile antennas, puts you at a disadvantage when competing with higher-power stations running larger antennas. That's especially true when you have a low-profile situation that limits the quality and effectiveness of the antennas you can use. But there

Figure 36 — Elecraft K2 transceiver for QRP operation. Note that you can install Elecraft's KPA100 option into the K2 cabinet to make it a model K2/100, with 100 W output. (Photo courtesy of Elecraft.)

You Have Been Bitten by the Powerful Flea-Watter of W8BAE

Early in my ham years, when I was living in Tennessee, I made an interesting contact on 40-meter CW with Ivo Depenbock, W8BAE, in Cincinnati, Ohio. Ivo used a teeny homebrew transmitter built around a 6SN7 dual-section tube in a crystal-oscillator/power-amplifier circuit. I use the common term "power amplifier" advisedly — Ivo's input power was 2 W. His output power might have been 1.5 W.

Over the next few years, I worked Ivo several times, and his signal was always good copy. We enjoyed some good rag-chews. Ivo was obviously having a ball with what we now call QRP operating. Quite frankly, I don't know if Ivo ever ran anything other than his "powerful flea-power 2-watter."

Those successful contacts with Ivo showed me that a ham could make good contacts with power outputs as little as 1.5 W output.

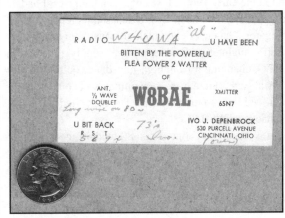

Figure 37 — The QSL card of W8BAE (SK) for a contact with his "powerful flea power 2 watter," a transmitter built around a single 6SN7 dual-triode tube.

is a solution: Improve your operating skills.

To encourage your thinking about improving your operating skills, carefully read the operating tips in Chapter 15 of this book. Find other books or magazine articles about operating, specifically QRP operating.

There have been many, many articles about QRP operation published in *QST* over the past 20 years or so. You can look them up to add to your store of knowledge about QRP and operating tips. The ARRL offers a number of books specifically about QRP operation as well as DVDs that contain scans of the issues of the magazine *QRP Quarterly* from 1979 to 2004. You can find a list of those QRP publications at **http://www.arrl.org/catalog/ ?category=&words=qrp&SearchWords.x=10&SearchWords.y=9**.

QRP operation can be a lot of fun. It's always a challenge to try to do things "the hard way." But you might be pleasantly surprised at how effective

QRP can be, even with the limitations placed on your ham station by having to use low-profile antennas. I suggest that you consider giving a try at QRP operation. See how well you can do. You might enjoy your QRP experiments enough to continue to use low power.

Back in the early days of my own ham operations, I used several transmitters. Some were homebrew, then on to an Eldico TR-75TV and three Johnson Viking rigs — Adventurer, Ranger, and Navigator. All those transmitters ran 75 W input or less (and perhaps 60 W output).

My first rig, which I used as a Novice and then as a General class licensee, was the 6AG7 crystal oscillator built from the two-part article by Don Mix, W1TS ("The Novice One-Tuber," *QST*, May and June 1951). That rig ran about 7 W input (5 W output). And I made contacts with it, even with my low level of experience. Therefore, I knew from my own experience that it was possible to make good contacts over fairly long distances with as little as 5 W output.

For a short period in the early 1960s, I was between ham rigs. I had sold my Central Electronics 20A but hadn't yet purchased its replacement, a Johnson Viking Navigator. I had a VFO on hand that was sold as a kit by Allied Radio. It was self-contained, with its own power supply. I made some modifications to the VFO, changing it from high-impedance output (to plug into a transmitter's crystal socket) to a pi-section output, so I could feed the low-impedance coaxial feed lines to my antennas. The converted VFO QRP rig ran about 2 W input (shades of W8BAE).

Using that little rig, I made good contacts on 80- and 40-meter CW over a period of several months. The hams I contacted would often ask me for a repeat of my transmitter's power when I told them my power — they found it difficult to believe that such a good signal could be coming from such a low-power rig, and thought they had perhaps miscopied my statement of transmitter power.

After my contacts with W8BAE, I never hesitated to use very low power for hamming. If you haven't tried QRP operation, you should. But let me warn you that you could get hooked on it, just as thousands on hams have become hooked on seeing what they can work with 5 W or less.

17

Low-Profile VHF/UHF Antennas

Because of the much shorter wavelengths, low-profile antennas for VHF and UHF are much easier to install than are low-profile antennas for HF. If you have access to your attic space, you can usually install very effective antennas there — rotary beams and the like. If your attic is large enough to give you room for the beam's rotation, you can even install long-boom Yagi-Uda beams.

In doing research for this book, I found some interesting approaches to low-profile VHF antennas that hams had used and described in *QST* articles. Here is a quick overview of some of them:

In "The Doctor is IN," in the March 2004 issue of *QST*, Jeff Graham, KR4ST, describes a small loop antenna for 2 meters that he built into a Hula Hoop. The matching system was made from a shorted 3½-inch stub (made from twin lead) and a 35 pF variable capacitor. The accompanying photo shows the antenna.

This Hula Hoop antenna could be used as a low-profile antenna by either installing it in your attic, or by extending it out a window (preferably after darkness has fallen) on a short wooden pole. This is a good example of what you can do in designing your own antennas, if you study antennas and learn how to build them and match them to the feed line.

Another interesting 2-meter antenna is the squalo ("square halo") that was described by Dick Stroud, W9SR, in his article, "Try Copper for 2 Meters — The Cu Loop," in the December 2002

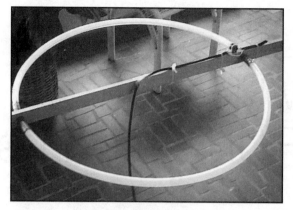

Figure 38 — KR4ST's hula hoop antenna for 2 meters. Just in case you have a hula hoop in your home that has fallen into disuse.

 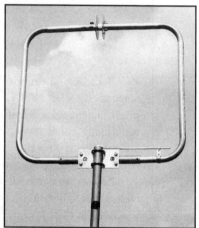

Figure 39 — Before and after photos of W9SR's aluminum-frame lawn chair that was turned into a 6-meter squalo.

issue of *QST*. That loop, made from plumber's thin-wall rigid copper tubing, can be mounted for either vertical or horizontal polarization, and can be used as a low-profile antenna as noted in the description of KR4ST's Hula Hoop loop.

Another fun antenna that W9SR built is described in his article, "Six Meters from Your Easy Chair," in the January 2002 issue of *QST*. Dick used the aluminum frame of an old webbed-seat lawn chair to build a 6-meter squalo. It is, indeed, a good antenna, and the idea of using an old lawn chair for the antenna materials tickles my funny bone!

Another 2-meter antenna that is quite practical, but which also amuses me, is one designed and built by John Purtune, W6NBC, and Fred Adams,

Figure 40 — W6NBC and WD6ACJ put their heads together to come up with this weathervane antenna. No one would recognize it as a 2-meter antenna.

WD6ACJ, "A Weather Vane Antenna for 2 Meters," which appeared in the August 2005 issue of *QST*. The accompanying photo shows the weathervane antenna in all its glory. In case you haven't figured it out, the supporting structure for the compass letters is a square loop that is tuned to 2 meters! I think John and Fred had a lot of fun building that antenna!

The discone antenna has been used over the past half-century by quite a few hams. The discone concept was first described in *The Proceedings of the IRE* in February 1946, and has been discussed in a number of variations since then in *CQ* and *QST*.

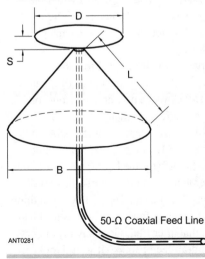

50-Ω Coaxial Feed Line

ANT0281

Figure 41 — The discone antenna can be built from sheet metal for the VHF/UHF range. Larger discones for HF can be made from wires or copper or brass rods as a skeleton discone. The dimensions L and D and the spacing S determine the frequency range covered by the antenna: L = 246/f (in MHz) at the discone's low-frequency end. D should be between 0.67 and 0.70 times the value of L. The spacing S between the disc and the cone should be between 2 and 12 inches — 2 inches for the smaller discones and 12 inches for the larger ones.

The first amateur publication of the discone antenna was in the article, "Discone — 40 to 500 Mc Skywire," by J. M. Boyer, W6UYH. Bob Patterson, K5DZE, presented a good article, "A VHF/UHF Discone Antenna," in the May 2003 issue of *QST*.

The discone's main advantage is that its frequency coverage usually spans a very wide frequency range, so that it can cover several amateur bands with a single antenna and single coaxial feed line. In addition to discone designs that were made from sheet metal discs and cones, other designs have used wires or rods to make a skeleton version of the discone. You can make a skeleton discone for VHF/UHF use and either install it in your attic, or carefully place it in a tree, so that it slips down inside the foliage of the tree. With the tree-concealed version, all you would have to do is to find a clever way to conceal the coaxial feed line.

The accompanying figure showing the basic discone design was taken from the 20th Edition of *The ARRL Antenna Book*. That publication has a good discussion of the discone antenna on pages 7-29 to 7-36 that I recommend you read, if you're considering a discone for your VHF/UHF station.

L. B. Cebik, W4RNL, came up with a good idea for a VHF/UHF antenna for amateur satellite use, a turnstile array for omnidirectional use using Moxon rectangles. See his article, "A Simple Fixed Antenna for VHF/UHF Satellite Work," in the August 2001 issue of *QST*. The low-profile ham could use such an antenna for satellite work from his home, if he has a good view toward the satellite from a window (or windows).

W4RNL also designed and built an antenna for 2 meters (and a little more), which he discussed in his article, "An LPDA for 2 Meters," in the October 2001 issue of *QST*. That log-periodic dipole array covers 130 to 170 MHz, and is well suited for low-profile hamming on VHF and UHF. You can hit the books and design your own LPDA for covering even a wider piece of the spectrum, in order to include additional amateur bands. For design information and construction tips on the LPDA antenna, refer to Chapter 10 of the 20th Edition of *The ARRL Antenna Book*.

Speaking of LPDA antennas… For two years, from 1962 to 1964, I worked as a civilian engineering consultant in Germany with units of the now-defunct US Army Security Agency. The LPDA was a new antenna concept at that time, and its wideband coverage was perfect for many of the ASA missions. I designed and built a number of LPDA antennas for the ASA units I supported, covering frequencies from the high HF range into UHF. I built one VHF/UHF LPDA antenna from brass tubing (for the boom) and brass rods (for the elements). I found a shop in Nuremburg (reminiscent of a medieval alchemist's shop) where I had the antenna silver plated. The LPDA was a thing of beauty, in addition to being a good performer. A few years later, after I had come back to the US, I heard via a friend that after I left Germany my silver-plated LPDA had been painted with olive drab Jeep paint so it would look more military for a big inspection. *Ouch!*

As the preceding discussion indicates, there are a lot of good antenna designs in the amateur literature that can be used directly or adapted for low-profile hamming on VHF and UHF. As I've mentioned before, keep your eyes open for good ideas that you can use for your own low-profile hamming. Now let's look at a few very simplistic antenna ideas for VHF and UHF that can get you on the air from your CC&R'd home.

First, let me get on one of my favorite soapboxes for just a moment.... *Please* don't try to get by with an indoor rubber ducky on 2-meter FM. Rubber duckies have been over-used and misused by too many 2-meter FMers who expect too much in the way of performance from them. You should consider the rubber ducky as an antenna for short-range simplex work, and for getting into repeaters that are within a few miles of you.

I'm sure you have heard a rubber-ducky user on a repeater who is breaking up so badly that you have difficulty understanding him or her. Please don't be one of those hams. It's easy to install a low-profile antenna for FM work on 2 meters and other bands. Please don't overuse your rubber ducky.

Soapbox — OFF.

Getting back to simplistic antennas, one very easy one that you can whip up with some stiff wire and a length of coax is a ground-plane antenna with two radials. For 2-meter FM use, cut one piece of the stiff wire to a length of 19¼ inches and another piece to a length of 38½ inches. Connect the braid of the coax feed line to the center of the 38½-inch wire and the center conductor of the coax to one end of the 19¼-inch piece of wire. Install an RF connector that will mate with your transceiver at the other end of the coax feed line.

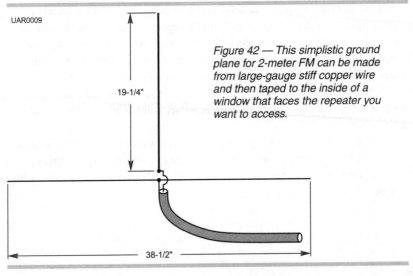

UAR0009

19-1/4"

Figure 42 — This simplistic ground plane for 2-meter FM can be made from large-gauge stiff copper wire and then taped to the inside of a window that faces the repeater you want to access.

38-1/2"

Tape the antenna to the inside of one of your windows, with the 38½-inch wire horizontal and the 19¼-inch wire vertical, above the horizontal wire. What do you know! It's a vertical quarter-wave vertical wire with a two-element counterpoise! If you tape this simple antenna to a window that faces in the direction of a repeater, you should be able to get into the repeater with a decent signal.

Another simplistic antenna to get you on 2 meter FM is the coaxial dipole. Strip the black plastic jacket from about 25 inches at the end of a piece of small-diameter coax (RG-58 or so) that has a woven braid shield. Place the tip of a pencil under the exposed woven-shield braid here and there, along its length, and gently work the inside diameter of the shield to be larger. Fold the braid of the coax back over the black plastic jacket until you have the exposed braid all folded back over the black jacket. Measure from the folded end of the braid (the center of the coaxial dipole) along the folded-back braid to the 19¼-inch point, and cut the braid off at that point. Place a small piece of black electrical tape around the end of the braid, to keep it from fraying.

Measure the exposed insulated center conductor of the coax (from the folded end of the braid to the end of the center wire) to be 21 inches in length. At the end of the center conductor, strip off the clear plastic insulation 19 inches from the center of the antenna (the point at which the braid is folded back). Form a small loop with the center conductor that is ¼-inch in diameter, trim off any excess length of wire, and gently tin the loop with solder.

You now have a coaxial dipole 38½-inches long. You can hang the dipole vertically, using the small mounting loop at the top end of the antenna, either by taping it to a windowpane, by hanging it from a push pin driven through the loop into the wooden window frame, or by hanging it from a short string to a convenient support.

If, when using simplistic antennas such as the ground-plane vertical or coaxial dipole, you are out of range from the repeater you wish to hit, you can build or buy a small beam antenna and mount it on a short wooden mast that is held upright by a tripod. Place the beam aimed out the window toward the repeater. That should get you on the air with a good signal. If you still have trouble hitting the repeater you want to operate through, you can add a small amplifier at the output of your transceiver to buy another 10 dB of transmitted signal.

I hope this discussion has given you some ideas of how you can get on VHF and UHF from your low-profile ham station. Keep reading the amateur literature for more ideas for antennas. Learn more about antennas. Design your own and see how well they work!

18

Taking Your Amateur Station into Deep Cover

By "deep cover," I mean the case where you would completely hide your identity as a ham from your neighbors. Quite a few low-profile hams do this, especially the hams who live in high-density housing such as large apartment buildings. The major reason for going into deep cover is to avoid dealing with neighbors who *imagine* interference problems from your station.

Many deep-cover hams have stations that are concealed inside furniture. Jim Kearman, KR1S, is the author of *Low Profile Amateur Radio*, the ARRL's first book on the subject. On several occasions in past years, Jim has gone into deep cover with his station. The accompanying photo shows Jim at his station in the early 1990s. As you see, the complete station is installed in an armoire. With the armoire closed, visitors wouldn't be aware that there's a ham station on the premises.

If you already have low-profile antennas, the only other things you have to do to go into deep cover is (1) to dress your feed lines inside your home so that they are completely hidden, and (2) hide the station equipment inside a piece of furniture (as KR1S did) or use a closet as a ham shack. With a closet as a ham shack, all you need do to hide the station is to close the closet door.

When hiding the ham station in either a piece of furniture or a closet, be sure and have adequate ventilation for the equipment (as well as the operator). Good old Wal-Mart has several models of

A Case of Imagined Ham Interference

In 1954, I was still in my teens, and active on HF CW. One day, a friendly neighbor (Ridley Parkerson, W4UQV, now a Silent Key) told me that another neighbor said I was causing interference to his TV set, that he had complained to me repeatedly, and that I refused to do anything about the interference. He also said he was going to complain to the FCC about the interference and my refusal to do anything about it. (Although you may find this hard to believe, back then the FCC dealt sternly with interference complaints.)

Funny, but that was the first I had heard of the interference — the neighbor had *never* said a word to me about it. I went to the neighbor's home, knocked on his door, and told him what I had heard. He was slightly flustered, but warmed up to the subject and soon started complaining heatedly about the interference I was causing "all the time" to his TV set. I asked that he turn on his TV and we started watching it for interference.

Within about three minutes, the interference started up. He said, "There! That's the interference you're causing!" I pointed out that I was standing there in his living room and not at my ham station and that my equipment at home was turned off. He again became flustered.

With his permission, I started checking electrical devices in his home, leaving the TV set turned on to watch for the interference, which continued in short bursts. Within about five minutes, I had located the problem. The "butter keeper" in his refrigerator was the culprit.

A butter keeper was the latest in refrigerator technology in the early 1950s. A small compartment in the refrigerator door provided a space where a butter dish could be stored. A small heating element in the butter keeper compartment kept the butter at a slightly higher temperature than that of the refrigerator, so the butter would cut and spread more easily. A thermostat cycled the heating element on and off fairly often, to maintain the desired temperature.

In my neighbor's refrigerator, every time the thermostat turned the butter keeper's heating element on or off, the thermostat contacts arced, causing the burst of interference as it switched. I explained my hypothesis to the neighbor. He unplugged the refrigerator as a test, we watched the TV for five minutes or so, and there was no interference. He plugged the refrigerator in again, and the interference immediately came back.

I suggested that he could have a refrigerator repairman correct the problem. Case closed.... And that, dear reader, serves as an example of why some low-profile hams choose to go into deep cover. People today are considerably more "off the wall" than they were 50 years ago. They are more likely to cause problems to the innocent ham who is not, in fact, causing interference Therefore, I can see why some hams resort to deep cover hamming.

Figure 43 — Jim Kearman, KR1S, had this deep-cover ham station one time in the past. The entire station was inside an armoire; when it was closed, no one would ever know there was a ham station on the premises.

Figure 44 — Jeff Mitchell, N1YDU, kept his deep-cover ham station in a small corner cabinet. (N1YDU photo.)

small fans priced around $5. A couple of those small fans placed in strategic locations can provide more than adequate cooling for your station equipment. Keep your eye open at hamfests for small box fans (either 12 V dc or 120 V ac models) — they are usually offered at bargain prices. Because the box fans are so compact, they can be placed behind equipment (where the heat sinks are usually located) more easily than can the Wal-Mart type fans.

19

Going Mobile as an Alternative to Fixed-Station Operation

If all else fails, and you are not able to come up with an HF antenna that will escape detection, you can operate mobile from your vehicle with considerable success and satisfaction over the long term. Here's why I believe that to be true.

In the 1980s, I was living in a rural corner of Montgomery County, Maryland. Our home's electrical service was by one of the older medium-voltage distribution lines. Generally speaking, electric power companies operate pretty much the same as any business nowadays — they try to get by with the minimum of routine maintenance. And that was certainly the case with the distribution line that brought electricity to our home.

When we moved into our home in 1972, I found that the RFI from the power line was amazingly high. It started in the AM broadcast band and was still strong on the high VHF TV channels (channels 7 to 13; 175.25 to 211.25 MHz). I was flabbergasted at the strength and the wide bandwidth of the RFI. On the HF bands, I could hear only the strongest signals — not a very good situation for working DX.

I called the power company repeatedly to report the problem. Finally, after about a dozen calls over a period of two months, a power company "RFI expert" came to my home. He had a small, battery-powered RF-sniffer radio that was about the size of the old Zenith Transoceanic radios.

The K3KMO Toyota-Van Mobile

For many years (back when my call sign was K3KMO), I operated mobile CW from a Toyota van, using a Ten-Tec Century 22 rig that delivered about 25 W output. At first, I used a Hustler mobile antenna, with several resonators in the van for changing bands.

Then I added a VP-1 fixture, which holds as many as four resonators on the Hustler mast. Electrically, this was a good antenna, but it was pretty top-heavy, which eventually caused mechanical problems. Soon I went on to my final mobile antenna system, one that worked quite well over the years.

I fabricated a large multi-antenna mag mount from a four-foot length of rectangular cross-section aluminum stock (a stock hardware store item). I bolted four aluminum cross pieces to the four-foot piece of aluminum, each a foot long. At the ends of those cross pieces I mounted eight large magnetic feet that I bought from one of the amateur mag mount manufacturers. I mounted five antenna mast receptacles on the mag mount, equally spaced along its four-foot length.

I connected separate runs of RG-58 coaxial cable to each antenna receptacle and ran the bundle of coax into the van, via the rear hatch, to an antenna switch. I also ran a heavy piece of ground braid from the mount to the body of the van, just inside the rear hatch. The reason the ground braid was required is interesting: When I first put the mag mount in place and tuned all the antennas to the desired frequencies, the antennas worked well from the RF standpoint. However, I was having electrical noise from the van's engine and ignition system that I had *not* had with the previous antenna system! Hmmmmm....

Because the noise was new, I reckoned that it was because of the new antenna system. I added a heavy-gauge, low-impedance ground strap to provide a good electrical and RF ground to the mag mount. The ground strap cured the RFI from the van, but adding it did not affect the tuning of the antennas.

The antennas I used with this big mag mount were *Hamsticks*. While tuning the individual antennas, I found that there was a lot of interaction among them because of the mutual coupling from each antenna to the other antennas. Eventually I learned that the easiest tune-up procedure was to tune the lowest-frequency Hamstick first and then work upward in frequency band as I tuned the other Hamsticks. I would tune all five antennas like that, then go through the tuning of the five antennas a second time to tweak them more exactly to the desired center frequencies.

Another thing that was necessary with this antenna system was to have quick-disconnect connectors at the base of each antenna. With

He turned on the radio in my driveway and tuned it around the HF and VHF segments of the spectrum ... and he couldn't hear any interference at all! The RF sniffer apparently had a very bad cold in its nose. So we went into my house and I showed him the interference on the TV screen and on my HF ham rig. Much mumbling and foot shuffling ensued.

After two more months of calls to the power company, a line crew finally those, I could take the antennas down if I had to go into a parking garage or if I was faced with other low-overhead situations.

That five-antenna system worked very well for me. There were some mechanical problems, but they were more amusing than annoying. Under some circumstances when driving along bumpy roads, the individual antennas would swing far enough fore and aft to start hitting their neighboring antennas — I called this the *sword fighting* effect. My home was in a forested area, and I would routinely drive under low-hanging branches of trees. The whips — all five of them — would strike the lower tree limbs and make quite a racket, with the van roof acting as a large sounding board. I got used to this noise and didn't think anything about it.

But one day as I was giving a friend a ride, we drove under some low-hanging tree branches, causing a big racket. The friend, accustomed to the frequent motor vehicle accidents in the DC area, literally jumped and ducked, thinking we were being hit by another car. After that, I always tried to remember to warn passengers there might be noise from the antennas.

Figure 45 — The serious HF mobile antenna system used by K3KMO on his Toyota van had five Hamstick mobile antennas, each fed with its own coaxial feed line, and a five-position antenna switch beside the Tec-Tec Century 22 transceiver. I worked for a Department of Defense contractor in the Maryland suburbs of Washington. One day at work my employer's Chief Security Officer called me and asked if that was my "white van out front with all the antennas on it." When I replied that it was, he told me that a visiting Navy office had reported the van to Security, thinking it might be the land-based version of a Soviet fishing trawler.

came out to inspect the wire wraps at the pole insulators on the offending line. Sure enough, they found quite a few older wire wraps that had become intermittent connections and were arcing, thus causing the interference. For the moment, my interference went away.

About two months later, the interference began again, caused by another group of wire-wrap connections that had become intermittent. After repeated phone calls to the power company over a period of about two months, they again sent out a line crew, which found a number of arcing connections and repaired them, curing the interference for the moment.

For about two years, that cycle was repeated. It would take two months of my complaint calls to get a line crew out and fix the wire wrap arcing, which would temporarily cure the interference — for about two months. That is, in effect, a 50% duty cycle of the interference over a long period.

Fortunately, the FCC will nowadays require that power companies *not* cause interference to hams (as they should have been doing all along). We can only hope that this enforcement of the FCC's regulations in this area will continue.

Because I enjoy chasing DX and making weak-signal contacts, this ongoing interference problem eventually annoyed me to the point of giving up home station operation and going mobile on HF CW. At the time, my commute to work was via a 25-mile path through rural countryside (corn fields and turf farms), with very little traffic. My typical driving time was 40 minutes. That was enough time for some good hamming. I used a multiband antenna system, so I could change bands from the driver's seat to follow good band conditions.

My results with my mobile station were excellent. Using a 25 W transceiver, I worked about 160 countries over the next three years, often making contacts through pileups. I had long rag-chews with friends, both old and new. On weekend trips I would often operate in contests. It was good hamming. Even on 80 meters, where the mobile antenna efficiency is in the order of 5%, I could work Europe with my 25 W when conditions were good. (See the sidebar in this chapter for a description and photo of the antenna system I used with my mobile station in my Toyota van.)

If you can't figure out a way to operate from your home, I recommend that you try mobile hamming. The equipment and antennas that are available today are excellent for mobile operation.

If I were starting from scratch today to get back into mobile operation, I would use one of the screwdriver antennas. I'd also have a way to tune the antenna without having to divert much attention from my driving duties. For example, Tarheel makes a programmable counter for their screwdriver antennas, which makes it easy to return to a specific frequency at the push of a button.

One interesting possibility presents itself to the CC&R'd ham who sets up a mobile station. If his vehicle can be parked near his home, he can place a run of coaxial cable underground from the parking place to the home, running the coax as much out of sight as possible at the home end. That coax could be connected to the mobile antenna and to the ham equipment in the home, enabling the ham to operate from his home station using the mobile antenna.

I have known hams who have done that with good success. If you use a system like that, you have to be certain that you disconnect the underground run of coax from the vehicle-mounted antenna before you drive away!

If you are considering going mobile, either as an alternative or in addition to your home-station operation, you should carefully read the ARRL book *Amateur Radio on the Move* for extensive information about mobile installations.

20

Concluding Remarks

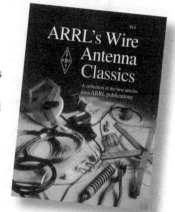

It is not within the scope of this book to provide a complete tutorial on antenna theory, construction, and adjustment. However, the more you know about antennas and how to make them play well, the better. Therefore, it is to your advantage to study antenna theory until you understand how (and why) antennas work the way they do, so you can figure out how to build an antenna, and how to adjust it to make it work as well as possible.

As an ARRL member since 1953, I've fallen into the habit of using ARRL publications as a major source of my Amateur Radio technical information. Therefore, I recommend the following ARRL publications (and RSGB publications available through ARRL Headquarters) to help you learn about antennas, how they work, what you can do to tweak an antenna until it performs as well as you want, etc:

◆ *The ARRL Antenna Book*
◆ *ARRL's Wire Antenna Classics*
◆ *More Wire Antenna Classics, Volume 2*
◆ *ARRL's VHF/UHF Antenna Classics*
◆ *Practical Wire Antennas*
◆ *Practical Wire Antennas 2*
◆ *Simple and Fun Antennas for Hams*
◆ *Backyard Antennas*
◆ *ARRL's Yagi Antenna Classics*

The Antenna File

- ◆ *The Antenna Experimenter's Guide*
- ◆ *International Antenna Collection Volume 1*
- ◆ *International Antenna Collection Volume 2*
- ◆ *HF Antenna Collection*
- ◆ *HF Antennas for All Locations*
- ◆ *Practical Wire Antennas*
- ◆ *Vertical Antenna Classics*
- ◆ *Physical Design of Yagi Antennas*

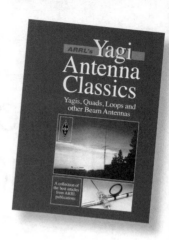

Antenna Topics

- ◆ *The ARRL Handbook for Radio Communications*
- ◆ *The Antenna Compendium* series (Volumes 1 through 7)

You might be particularly interested in the seven volumes of the popular *Antenna Compendium* series, published between 1985 and the present. Each volume presents a great number of antenna articles that were not published in *QST*, so you may not have seen them. All volumes of the *Antenna Compendium* series are available for purchase from ARRL, with the exception of Volume 4, which is out of print.

Go to the ARRL Web page **www.arrl.org/catalog/** for ordering information on publications and products available through the ARRL. In addition to the antenna-related publications, check out the other ARRL publications and articles — you might find some things you can't possibly do without.

Also, you might like to read more about QRP operation in the ARRL book, *Low Power Communication.*

The above list is not intended to exclude the many antenna publications that are available from other publishers, so check the availability of other literature in the field, to include articles in the ham publications (both current and past) as well as antenna books.

You might not be able to duplicate an antenna described in the amateur literature because of the physical limitations placed on you to maintain your low profile. Nevertheless, those publications are good background reading to help your thinking when you start designing an antenna to fit your particular limitations.

SAFETY ISSUES

You should always keep in mind that playing with radios can be dangerous. You are exposed to the possibility of electrical shock or electrocution, and to falls from towers, rooftops and balconies while working on antennas. You are required by the FCC to assess the strength of RF fields in and around your ham shack.

Rather than rehash the details of the safety issues you should consider, let me instead direct you to other publications for further reading. Read the chapters on safety in *The ARRL Handbook for Radio Communications* and *The ARRL Antenna Book*. Read the ARRL book *RF Exposure and You* for information on RF radiation safety issues.

As an example of how you can get bitten when you least expect it, here's something I did in the late 1960s. I was working around a 6 V dc auto battery. Who can get shocked with 6 V, right? As I moved my hand to reach something, my wedding ring came into simultaneous contact with the positive battery terminal and ground. I didn't get shocked, but enough current flowed through the ring to heat it to a *very* high temperature. *Ouch!* Even though I removed the ring from my finger as quickly as possible, within an hour I had a big blister all the way around my finger where I wore the ring. After that, I always removed my ring before working on anything electric!

Just remember the ABC of Amateur Radio — Always Be Careful!

A FINAL WORD

Don't be discouraged if you are in a home with CC&R or other limitations to your hamming. Even small antennas can allow you to have plenty of good contacts. Put up some wires. Match them to your rig. Start hamming it up. As Nike says, "Just do it!"

Appendices

The following appendices present reformatted articles from *QST* that are especially pertinent to the subject of low-profile HF antennas. If you have access to a large collection of back issues of *QST* or the *QST View* CD-ROMs (available from the ARRL in 13 sets that cover the issues of *QST* from 1915 through 2004), you can browse through the back issues at your leisure to look for antenna articles that you might be able to use in your low-profile station.

But here's a fair warning: Once you start looking through those back issues of *QST*, you will quite likely spend time looking at other articles as well as antenna articles. There's lots of good information and interesting reading in those back issues!

Appendix

The following pages are reprinted from the November 2005 issue of *QST*. Bruce Pontius, NØADL, discusses the simplistic methods he used to hide his wire antenna from the view of his neighbors, and the good results he has obtained with it.

In Figure 2, note the cheat sheet Bruce made of his antenna tuner settings. That chart enables him to change bands and quickly return to the settings of the controls on the antenna tuner. Also note Bruce's comments on "Developing an Effective Counterpoise."

Surprising Results with a Low, Hidden Wire

Here's how one amateur made a stealth wire antenna system that doesn't give up much performance.

Bruce Pontius, NØADL

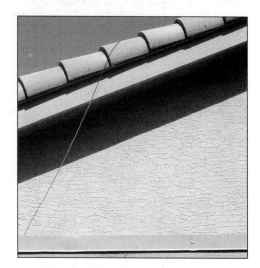

We live in a community that does not appreciate out-door antennas of any kind. In addition, the houses are close together and RF interference can be a problem at neighbors' houses as well as mine. Occasional minor interference seems less objectionable to neighbors if they don't see antennas to remind them of (or alert them to) the radio operations. I have good relationships with my neighbors, but I would rather avoid RFI discussions.

The 45 foot high horizontal loop and balanced feed system described a couple of years ago in *QST*[1] would be terrific, but I don't have supports to get a loop up in the air. Forget that! Like it or not, I have to use hidden antennas except for temporary operations. In this article, I discuss my experiences with an almost unnoticeable end-fed random wire HF antenna and counterpoise system. I tend to use short transmissions on SSB and low power for digital modes, and I have had many rewarding and enjoyable contacts with this simple setup. If you're in a neighborhood that's unfriendly towards outdoor antennas, perhaps you will be encouraged to try your own stealth antenna.

"DESIGN" OF THE END-FED RANDOM WIRE ANTENNA

I started with a chart of relative impedance versus wire length for the HF bands. I chose a length of 92 feet, which provides a reasonable impedance for an antenna tuner on most bands. At 15 meters, the wire is nearly a multiple of ½ λ, presenting a very high impedance at the antenna tuner. This might also be the case at 60 and 30 meters, but these two bands are not a main goal for me. After building the antenna, I was pleasantly surprised that my MFJ-949D tuner could provide a match to the transmitter at 15 meters and does not arc or misbehave, at least at 100 W input. The antenna is about ⅜ λ at 75 meters, and it works there, at least a little.

The next task was running the wire outside and along the roof of my house. Most of the antenna is 20 gauge insulated wire, with 18 gauge stranded bare copper to lower visibility where the wire is above the roofline and in silhouette. The wire starts at the operating location and passes through an exterior wall just above the patio roof behind the house. Then the wire runs up to the main roof, following a tortuous path over the edge of the tile roof, up to a low chimney, and then along the peak of the roof. It is lying on the tile for much of its length and in some places is held slightly above the roof where it is tied to the chimney and, further along, to a vent pipe. The last 6 feet or so of the wire, at the open end, is held off of the roof tiles by finding appropriate places to tie it. The highest spot is at the chimney, 17 feet, 9 inches above the earth.

To get the wire outside, I drilled a ⁵⁄₁₆ inch hole through the wall from the outside and drew the wire through the wall. Be sure there are no wires or pipes where you drill! The holes were plugged with caulking material. If the wire is removed, the holes can be filled and painted to match. Inside, the entry point is concealed by a china cabinet and the wire can be hidden when not in use. As indicated in the lead photo, the hidden wire antenna is barely noticeable as it exits above the patio roof and disappears over the edge of the tile roof.

I use the antenna with a portable station that takes only a few minutes to set up and put away, leaving almost no traces of a radio station. Another radio operating position, in the middle of the house, has more permanent antenna feeds for indoor antennas or for temporary VHF/UHF or satellite operations.

DEVELOPING AN EFFECTIVE COUNTERPOISE

When I first tried transmitting with the random wire, there was RF all over the place, including a tingle in my fingers and nose. Adding wires to the ground lug on the antenna tuner to form a counterpoise improved things considerably. My counterpoise wires are made from insulated 20 gauge wire and tied with nylon string to furniture in adjoining rooms. The wires are 25-30 inches above the floor and close to ¼ λ at the frequency in use.

My first counterpoise didn't solve all the issues with stray RF. For example, on 40 meters, with only one ¼ λ radial, a ground fault circuit breaker in the adjacent kitchen picks up RF energy and trips. I started experimenting with the MFJ-931 Artificial Ground counterpoise tuner. With a second 40-meter counterpoise wire attached to the MFJ-931 and another '931 tuning a wire to an 8 foot ground rod just outside the wall, things are much cooler. There is no noticeable RF feedback or mischief except for the inevitable pickup in the wireless phones and audio equipment. The higher bands pose fewer ground system difficulties and do not require a tuner or a ground rod.

Figure 1 shows the system I ended up with. The transceiver is on the far right and the MFJ-949D antenna tuner is the lower box on the left. The red antenna wire goes up and to the left from the WIRE terminal on the tuner. Fixed counterpoise wires (one ¼ λ wire for each band in use) attach to the ground lug. More counterpoise wires, perhaps passing out through windows, might help performance but the complication does not seem worth the possible improvement.

One MFJ-931 counterpoise tuner is on the table between the transceiver and

Figure 1 — The station equipment setup, viewed from the back.

Table 1
Antenna and Radial Wire Lengths

Band (meters)	Center Freq (MHz)	Calculated ¼ λ (feet)	Length of Counterpoises (feet)	Vertical Radiator Length (feet)	¼ λ Radial Length (feet)
75	3.925	59.6	Ground rod		
60	5.350	43.7			
40	7.265	32.2	34.0, *32.4**	32.9**	32.9
30	10.125	23.1			
20	14.290	16.4	16.2, *14.6**	16.7	17.0
17	18.140	12.9	13.4, *14.6**	13.4	13.4
15	21.375	10.9	11.5, *10.2**	13.4***	11.2
12	24.960	9.4	9.4, *7.5**		
10	28.490	8.2	8.3, *7.0**		

*The length shown in italics is for an additional counterpoise wire tuned through the MFJ-931 Artificial Ground counterpoise tuner for operation on 40-10 meters (no counterpoise wire is used on 75 meters). Also, on 40 meters and 75 meters, another MFJ-931 tunes a wire connected to an 8 foot ground rod.

**This dimension was about 10 inches too short.

***The 17 meter vertical radiator was also used for 15 meters and the antenna fed through the tuner.

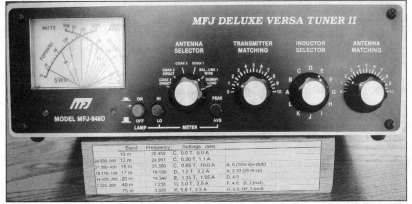

Figure 2—Antenna tuner settings can be recorded on a chart for faster band changes.

antenna tuner. The wire connected to the red terminal is approximately ¼ λ for the band currently in use, and I change wires and tuner settings when I change bands. Another counterpoise tuner is on top of the antenna tuner, with a wire going outside to an 8 foot ground rod. The ground rod alone is used for 75 meter operation because of the difficulty in routing effective ¼ λ (63 foot) counterpoise wires around the house. Once the system is tuned up, bandwidth on 40 meters and up is adequate to cover most of the General class phone sections with little or no retuning. Table 1 shows the wire lengths I used for the counterpoise system. The last two columns are dimensions for a ¼ λ vertical antenna used for comparison testing as described later.

This setup might seem complicated but it really only takes about 12 minutes to assemble the equipment and wires from their stowed positions for a day of operating, and then only a couple of minutes to change bands. To make band changes easier, you can make a chart like the one shown in Figure 2 to preset the antenna tuner and main counterpoise tuner.

Most of the transmitter power seems to be going into the antenna. The tuner has been tested with loads simulating the wire antenna. Losses appear to be less than 1 dB on most bands, but I have not checked 80 meters. The transfer function on all bands exhibits expected changes versus frequency, demonstrating that it is not just a flat-frequency-response "dummy load" antenna and matching system.

When using two or more counterpoise tuners with RF current sensors and meters, the effects of resonance in the counterpoise wires is quite evident. For example, I tuned one counterpoise for maximum RF current while transmitting (it's tuned for minimum RF impedance—what we want is a short at the tuner). When another counterpoise wire is added, the two wires are interactive. If one is detuned from resonance, the current increases in the other and vice versa. Probably the best situation is to roughly balance the two currents, or in the case of many counterpoise wires, to cause them all to be at resonance and to equally share the RF ground currents.

ON THE AIR WITH THE WIRE ANTENNA

I used a 100 W transceiver for the operating described here. Some contacts were made in 2002 and 2003, but most of my operating with this wire took place in February and March of 2004 after the sunspot cycle had continued its drop. The relatively small amount of operating time has resulted in Worked All States on 20 meters with QSL cards in hand. I'm just a few states short of Worked All States on 17 and 15 meters, and have Worked All Continents. A few days of operation thus far on 40 meters has yielded more than 30 states.

In only a few hours of operation during the 2004 ARRL DX Contest, I logged many stations on 10 meters and 15 meters, including all continents except Africa. My time only allowed a few contacts on 20 meters, as I concentrated on 10 and 15 while those bands were active.

The results that weekend encouraged me to try the CQ WPX contest at the end of March 2004. Again, I went for the higher bands first and logged more than 30 contacts on 10 meters and about 50 contacts on 15 meters in just a few hours of operating time. Contacts included DX stations on five continents. Running out of time, I reluctantly moved down to 20 meters and, of course, could have filled a computer log book with so many stations on the air. I was not able to make contacts through some pileups, but was quite satisfied with the percentage of successful attempts. I did listen far more than transmit. The activity on 40 meters seemed lower, at least during daytime, so I only tried for a while, logging a dozen or so in a half hour.

COMPARISONS WITH OTHER ANTENNAS

The performance of the hidden wire was much better than expected, so I invested some effort in on-site comparisons with other antennas to get an estimate of just how much performance must be given up in the use of the compromise antenna and ground system. It happened that both of my next door neighbors were to be gone for two weeks at the same time, so I could put up some big antennas and blast away without breaking my cover.

First, I set up a Force 12 Sigma 5 vertical dipole for 20-10 meters in the backyard with the lower end about 2.5 feet above ground. On-the-air transmit and receive comparisons showed that the wire often seemed to work as well or better on 20, 17 and 15 meters. I did not try 10 meters. The two antennas could be switched rapidly back and forth using the Kenwood TS-570D's two antenna ports and the front panel switch. The wire was fed through a tuner, but the Sigma 5 did not require one. I have heard of others' successful use of the Sigma 5, so I was pleased that the wire compared well with it. Therefore I decided to put in more effort to make comparisons with standard antennas.

BUILDING A COMPARISON ANTENNA

A standard half-wave, horizontal dipole would have been most desirable for comparisons, but I don't have sufficient supports. A vertical ground plane with elevated radials seemed to be the next best standard antenna type, and it was something that I could manage in my yard.

I considered buying a good vertical antenna but didn't want to complicate the comparison with a multiband unit. I wanted to have a basic, inexpensive $1/4 \lambda$ piece of metal for simplicity of measurements on one band at a time. Besides, I love the magic of fundamental electrical phenomena and looked forward to observing the interactions of simple vertical conductors and an elevated, tuned radial system.

The first vertical, for 20 meters, is shown in Figure 3. The vertical radiator uses an 8 foot length of 1 inch tubing with an 8 foot length of ⅞ inch tubing telescoped inside. A ⅜ inch aluminum rod clamped to the ⅞ inch tube brings the total length to 16 feet, 9 inches. The antenna base is 5 feet above ground, mounted on a 2-inch-square wooden pole with two U bolts as shown in Figure 4. The top of the pole is guyed with four nylon ropes.

Nine 17 foot radials (seven good ones and two bent to fit my yard) were fastened at the feed point by twisting the 18 gauge stranded copper radial wires around a hose clamp as seen in Figure 4. The radials are about 5 feet above ground at the antenna base and are stretched out and tied off with nylon string to stakes (or to anything handy).

The feed point connects with straps to the vertical radiator as shown in Figure 4. One strap is fastened with a screw and nut to the bottom antenna section and the other is clamped between the pole and antenna tube. An insulator can be placed on the wood pole behind the strap if desired. The straps are soldered to the connector center pin.

ADJUSTING THE VERTICAL

Before attaching all the radials, I adjusted the first four with the aid of instruments. Next I adjusted the vertical radiator for resonance at the desired frequency. Then I cut the rest of the radials to length, attached nine and again measured the antenna resonant frequency. Leaving the vertical radiator length alone at that point, I readjusted the radials to bring the antenna to resonance. This iterative process could be repeated, but results were good enough after one cycle. The final dimensions are shown in Table 1.

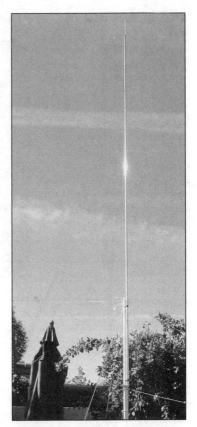

Figure 3—The 20 meter vertical antenna is about 22 feet tall, including the 5 foot wooden support post. It definitely attracts more attention than the hidden wire.

The signal source for these adjustments is similar to the one I described in *QEX*,[2] followed by a Motorola 2832C balanced, push-pull amplifier and a 3 dB attenuator. The power level through the precision directional couplers to the antenna can be less than 100 mW since sensitive power meters are used.

Thankfully, all this measuring must be done only once. After the antenna is adjusted, the SWR will be less than 1.4:1. The vertical can be set up anywhere and things will work properly unless large conducting objects are in close proximity.

I used similar procedures for the 17 and 15 meter versions. The 17 meter antenna used nine radials. The 15 meter antenna used all of the 17 meter radials, plus 6 more cut to the length shown and fit in between. I used the 17 meter vertical radiator length on 15 meters as well, and fed the antenna through a tuner.

Figure 4—Radials are wrapped around a hose clamp and pressed against the antenna connector. A brass strap is soldered to the connector center conductor and then bolted to the vertical radiator. Two U bolts hold the vertical radiator to the mast.

For 40 meters, I added another 8 foot section of aluminum tubing at the bottom and lengthened the ⅜ inch rod to make a 33 foot vertical. This antenna was guyed at the middle with ropes. The 40 meter radials were a compromise. Three 34 foot radials fit in the yard without bending, but three additional wires required bending to fit, with two of them quite contorted.

COMPROMISE WIRE VERSUS FULL SIZE VERTICAL

I compared the vertical and the random wire on 20 meters first. An operator in the Philippines thought the vertical slightly better, but at my end, on receive, the signal-to-noise (S/N) ratio was better on the (mostly horizontal) wire. Contacts in California favored the vertical, while contacts in Idaho and Washington favored the wire, as did local Arizona contacts. A couple of stations in Texas liked the vertical a little better, while stations in the Southeastern US gave better reports on the wire. The Upper Midwest and Northeast were a toss-up.

Received signals exhibited a better S/N ratio on the wire 75% of the time or more, but often the signal strength was somewhat higher on the vertical. Ignition noise was much louder on the vertical, but it can be reduced with the noise blanker and doesn't often bother me. Sometimes the signal strength and quality advantage at both ends of the contact shifted back and forth between the wire and vertical, adding to the conclusion that the two antennas performed about the same.

On 17 meters, a fair amount of listening revealed that the wire was definitely better on receive. As on 20 meters, the S/N ratio was better on the wire. Contacts in Mexico and the Caribbean favored the wire. Stations in South America and Japan reported equal performance. During a contact with a Texas station signals shifted back and forth, with one antenna better and then the other. A station in Alabama on a vertical antenna reported better signals from the wire. In general, the southeastern US seemed about the same on either antenna, as did the Midwest and New York. This was far from a comprehensive test, but it appeared that performance was similar to 20 meters with a slight advantage going to the wire.

After listening for a while, I didn't hear anyone on 15 meters so I resorted to calling CQ. Up popped VP6MW, one of 42 residents of Pitcairn Island in the South Pacific. We had a great QSO during which signals varied widely. The wire seemed to give a better S/N signal on receive, and VP6MW reported my signals about the same on the two antennas. During operations on a net, where protocol prohibited back-and-forth transmit comparisons, the antennas seemed to work about the same.

On 40 meters, the 34 foot aluminum vertical outperformed the wire on transmit, significantly in some cases, during contacts with 20 or more stations from around the US. Most reported stronger signals with the vertical, typically at least 1 S-unit and up to 2 S-units. As on the other bands, better receive S/N ratios on the wire helped the receive performance. On the lower bands, the wire is clearly a compromise as a transmitting antenna. Nevertheless, with limited time and effort, operation on 40 meters with the wire has yielded solid contacts in more than 30 states. More counterpoise wires might help.

THE WIRE WORKS OKAY, BUT IS IT SAFE?

RF safety is a concern with any indoor antenna or antenna close to the house. I was able to make some measurements of the electric fields of the hidden wire and comparison vertical antennas. Using these relative voltage measurements and information and tables from the ARRL publication *RF Exposure and You*,[3] I determined that the maximum permissible exposure (MPE) would not be exceeded using 100 W on SSB and short transmission times.

I based my conclusions on the tables starting on pages 8.26 and 8.37. These tables are for ground-mounted verticals and for elevated vertical ground planes. The worst case exposure situation for the vertical antennas is when the operator is sitting right within the radials field of an elevated vertical, among the radial wires, with head and shoulders above the plane of the radials. The tables say a person could be within 11 feet of an elevated groundplane vertical antenna and be under the exposure limit, even with more than 100 W PEP of transmit power. (Duty cycle and on/off times are considered and factored in for my type of operations on SSB.)

The wire antenna slopes away from the operator, with limited counterpoise wires extending away from the operator. It was assumed to exhibit lower energy fields near the base than the vertical antennas in the tables. Also, the total energy radiated in the near field from the wire is distributed somewhat along its 92 foot length. Considerably more than100 W of power could probably be used with the wire while remaining within the recommended MPE limits, however. Each operator is responsible for his own evaluation of his particular situation. *Note:* While the author's configuration may not exceed published limits, an antenna like this is hard to evaluate for RF safety, especially since so much of it is in proximity to people. To be safe, consider limiting power output to 50 W or less to be below the threshold at which measurements or analysis is required.—*Ed.*

CONCLUSIONS

The stealth wire works. Using an end-fed wire eliminates the need for a supported center feed, which can be difficult to do while keeping the antenna out of sight. If you have a problem with erecting a huge, high antenna, go ahead and string up this compromise. Even though it is a compromise, it is a quite effective antenna and counterpoise system—even as a portable station.

Notes

[1]K. Kleinschmidt, "A Balanced, Everyday Approach to All-Band Bliss," *QST*, Apr 2002, pp 47-50.

[2]B. Pontius, "Signal Sources," *QEX*, Nov/Dec 1999, pp 18-30. See Note 11 in this article.

[3]E. Hare, *RF Exposure and You* (Newington: 1998). Available from your local dealer, or

from the ARRL Bookstore, ARRL order no. 6621. Telephone toll-free in the US 888-277-5289, or 860-594-0355, fax 860-594-0303; **www.arrl.org/shop/**; **pubsales@arrl.org**. You can also find RF exposure resources on-line at **www.arrl.org/tis/info/rfexpose.html**.

Additional Reading

The author's original, unedited article with data and pictures is available on **www.radioadv.com**.

B. Muscolino, "My Antenna is a Compromise—and It Works!" *QST*, Apr 2003, pp 59-61. This article supports the use of simple end-fed wires, but does not provide much actual same-site comparison.

K. Kleinschmidt, *Stealth Amateur Radio* (Newington: 1998). This book is devoted to operating without calling attention to yourself. It's currently out of print but may be available on-line or at hamfests.

The ARRL Web site has information on limited space antennas at **www.arrl.org/tis/info/limited.html**.

Photos by the author.

Bruce Pontius, NØADL, holds a BSEE and has been involved in the development of semiconductors and radio equipment and systems for many years. He played major roles in the development of early cellular radio equipment, digital trunking radios and narrowband data radio equipment. Bruce served as Engineering Vice President at EF Johnson Company for 15 years and worked with other companies in similar roles. He now serves as president of TRM Associates and is working in wireless communications and RFID. Bruce first got involved with Amateur Radio at age 11, building radios and test equipment with his father. He has been licensed since 1978 and enjoys operating with simple equipment and portable operation. You can reach Bruce at 15802 N 50th St, Scottsdale, AZ 85254, e-mail **bepontius@aol.com**.

B

Appendix

The following pages are reprinted from the March 2005 issue of *QST*. Kim Stenson, W4KVS, tells how he uses simple techniques to put his antennas into a typical small attic, with remakable results on 20 to 6 meters.

On the HF bands, Kim has worked over 257 countries. He obviously has good ears and uses good operating and DXing techniques to do that well. Kim is a Poster Boy for low-profile antenna operation!

Keep in mind that you can emulate Kim's low-profile hamming. You can sharpen your operating skills, install some effective low-profile antennas, and have great fun on the ham bands.

Adventures with Indoor Antennas

W4KVS has a modest, inconspicuous and effective aerial arsenal—in the attic.

By Kim Stenson, W4KVS

A few days after receiving my ham license I purchased an ICOM IC-551D 6 meter transceiver (circa 1987), power supply, tuner and 6 meter dipole. Living in an antenna-restricted area I was forced to use an indoor antenna. I temporarily installed the dipole on the second floor of my two-story brick home and conducted an RF exposure safety check. I soon located a local net, the Sandlapper 6 meter Net. Not being sure my temporary indoor dipole would work, I made a call. To my surprise, K4IRT came back to me and I checked into the net. Over the next few weeks, I was also able to check into nets in both Georgia and North Carolina. Received signal reports were not good, but I was able to make contact.

IMPROVING THE INDOOR ANTENNA

In search of better performance, I moved the dipole from the second floor to a more permanent setup in the attic. Mounting the antenna there served three purposes: First, it would add at least 10 feet to the antenna height, putting it about 25 feet up. Second, it would reduce RF exposure. Third, my signal would only have to go through plywood and shingles. The available attic space was 15×25 feet but it had some drawbacks. Most of the attic was not decked, making movement difficult. Additionally, a large HVAC unit was located on what decking there was.

Working space was one thing, but I had no idea how the HVAC unit might affect operations. Despite the heating unit and the wood and shingles, signals improved significantly. Nevertheless, the search for a better system continued. After a little research, I determined a loop antenna might be a possible solution. I purchased and installed an M² Products 6 meter HO loop in the attic, approximately 30 feet above ground level. A view of the loop can be seen in Figure 1.

Local contacts immediately improved and shortly thereafter I experienced my first 6 meter opening and worked WA2SPL/1 in Vermont. Over the next several

Figure 1—The 6 meter loop mounted near the top of the roof.

weeks I worked numerous stations including K7MI in Oregon and HC1BI in Ecuador. During these openings I was able to compare the dipole and the loop in the receive mode. There was a noticeable 2 or 3 S-unit improvement using the loop (the M^2 Products HO 6 meter loop gain is advertised at 4.2 dBd at 11 feet above ground). Despite the improvement with the loop, I looked into the possibility of a beam in the attic—settling on a Cushcraft 3 element beam with a 6 foot turning radius.

Using a piece of metal pipe and a flange, I built a base for the antenna and installed it on one of the cross-beams. With a TV rotator and short mast, the beam itself was a little less than 25 feet above ground. Assembling all the antenna parts in the attic was a bit awkward, but successful. The rotator installed on a ceiling joist is shown in Figure 2 and a view of the beam can be seen in Figure 3.

Figure 2—The rotator is mounted to a pipe support secured to a floor member. Ceiling insulation covers the support.

Signals also improved with the use of the beam, but not as much as the loop over the dipole. With both antennas it was necessary to use a 6 meter tuner to obtain an acceptable SWR (probably not necessary outside, but a necessity in the attic).

Although I prefer the loop because of its omnidirectional capabilities, the beam is often handy in pulling out the weaker signals. Another advantage of the loop is that it is stationary and tuner adjustments are usually not required. The beam moves

Figure 3—The 3 element, 6 meter beam with crossed dipoles above.

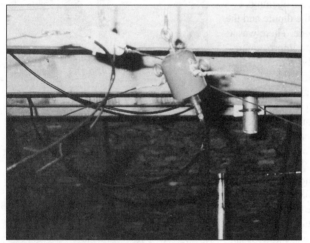

Figure 4—The 10 meter dipole center feed. Part of the second, orthogonal dipole can be seen above it.

and with it goes the SWR, which requires tuner adjustment when changing directions. Both proved to be acceptable indoor antennas. In less than a year on the air, I was able to work 10 countries, 40 states and over 100 grids on 6 meters.

ON TO HF

Exploration of the HF bands using an ICOM IC-718 was my next goal. A 10 meter beam seemed impractical so I settled on a dipole, whose center feed is shown in Figure 4. VHF operation with a dipole was a bit spotty, at best. I wasn't overly optimistic. The result, however, was pleasing, as the band was full of signals and I jumped right in.

The RF Proximity Question

Any time one operates with indoor antennas, consideration should always be given to the dangers of RF radiation. An attic antenna could pose a radiation hazard if the operator or, for that matter, anyone, is subject to RF radiation levels that are unsafe. How do we determine a safe radiation level? There are certainly standards, but these require sophisticated RF measurement instruments to get quantitative results. Fortunately, sophisticated measurement techniques aren't necessary if we keep the power levels *reasonable*. Much of the hard work has been done for us by means of accurate modeling software and representative antennas have already been modeled. (See *RF Exposure and You* by Ed Hare, W1RFI.[1]) A reasonable power level and one that will ensure compliance with accepted standards for an attic antenna system such as this is frequency and antenna gain dependent.

Consider a 10 meter half-wave dipole at a height of 20 feet (typical for an attic antenna in a 2-story home). Compliance is achieved if we limit our power to 50 W output and we are at least 3 feet away from the antenna. If we run 100 W output, we must maintain at least a 4 foot distance. These are so-called *controlled* limits and apply to the operator. If we consider *uncontrolled* limits (the general population), we must be 9 feet from the antenna for compliance at 100 W or 6.5 feet away for compliance at 50 W. Common sense tells us that we are indeed better off at low power levels. If we consider a 17 meter dipole at the same height, we must be 1.5 feet away at 50 W (controlled exposure) or 3.5 feet away (uncontrolled). For the same 17 meter dipole at the same height (20 feet) at 100 W, we must be 2.5 feet away (controlled) and 5 feet away (uncontrolled).

The 6 meter, 3 element beam is another story. For an attic antenna, if we are running 50 W, we must be at least 6 feet away (controlled) or 12 feet away (uncontrolled). If we lower our power to 10 W, we can be 4 feet away (controlled) or 6 feet away (uncontrolled). Clearly, it would be prudent to avoid antennas with meaningful gains in attic installations.

In all of these situations, common sense should prevail. Avoid high power levels and keep your distance. Also, remember the RFI (radio frequency interference) potentials as well. A by-product of the lower power levels will be lower tendencies for interference with common household appliances that are much closer to the antenna in an attic installation. Attic antennas certainly require compromises. Have fun, but keep the safety issues in mind.—*Stu Cohen, N1SC, Technical Editor, QST*

[1]Available from your local dealer or the ARRL Bookstore. ARRL order no. 6621.

My first HF contact was in Barbados, 8P9JW, who gave me a 59 report. Not bad, but the Caribbean is pretty close. I heard a QRP station in Colorado calling CQ. I adjusted my output to 5 W and answered. To my surprise KF7MD came back with a 56 report. Not as good as my report from Barbados but not bad for two way QRP. By this time it was late in the afternoon and 10 meters started to close. I couldn't wait until the next morning to see how I might do with signals coming in from Europe.

Early in the morning the band started to open up again with many signals. I heard Ted, LZ1WR, calling CQ. I called; he answered with a 59 report. The attic dipole was going to work!

After a couple of months working on 10 meters, I wanted to explore other bands. I extended my dipole to 12 meters by adding a foot (or so) of wire to each end of

Figure 5—The 10 meter dipole is lengthened for other bands with a simple extension on both ends. An alligator clip secures the extended element.

the dipole with an alligator clip. One end of this can be seen in Figure 5. The result was a sort of inverted U dipole. I experienced similar results with this setup and found I could easily work many stations. I even worked my first Japanese station on 12 meters. One drawback to this setup was that changing bands required a trip to the attic.

I decided to further modify my dipole and add extensions (again using alligator clips) to create a 20 meter zig-zag dipole. Results of this setup were also acceptable and I could easily work everything between 10 and 20 meters using a tuner. On occasion, I was even able to bust pileups with the dipole, including a memorable QSO with a station in Tanzania on 17 meters. I later installed another dipole at right angles to the original dipole with some interesting results. Signals on 10 through 17 meters were about the same, but the difference on 20 meters was amazing. Signals barely registering 1 unit on the S meter with the first dipole now boomed in at S9 with the second dipole. With my indoor dipole, loop and beam I was able to work stations in over 240 countries in less than a year of operation.

MAXIMIZE SUCCESS

Here are a few tips for a successful indoor antenna. Any antenna should be mounted as high as possible. A low dipole in the clear is not going to be particularly effective and will be less effective if under a roof at the same height. Place the antenna as far away as possible from any wiring or other metallic objects.

Expect at least some RF interference from appliances in the home and also expect to create some yourself. Plan on doing some work to mitigate that interference.

Indoor antennas can and do work well.[1] Are they comparable to outside antennas? Maybe not, but indoor antennas can provide many good contacts and a fair amount of DX.

Photos by the author.

Kim Stenson, W4KVS, was first licensed in 2000 and has worked 257 countries (250 confirmed) with indoor antennas. He earned a BA from Washington and Lee University and an MA from Norwich University. Kim is a retired US Army Infantry Officer and saw combat service during the (first) Gulf War. Currently, Kim is manager of National Hazard Plans, South Carolina Emergency Management Division. He may be contacted at w4kvs@arrl.net.

---

[1]*Stealth Amateur Radio* by Kirk A. Kleinschmidt, NT0Z , has additional information on invisible and indoor antennas. It is available from your local dealer or the ARRL Bookstore. ARRL order no. 7571. Telephone toll-free in the US 888-277-5289, or 860-594-0355, fax 860-594-0303; **www. arrl.org/shop/; pubsales@arrl.org**

C

Appendix

The following pages are reprinted from the January 2004 issue of *QST*. In this article, Rod Vlach, NNØTT, tells how he has learned to compete with the big guns on the ham bands, even though he has a "little pistol" station.

Rod isn't limited to low-profile antennas; nevertheless, he uses only modest antennas from a small city lot. His tips on operating techniques — including some sneaky ones — are required reading for the low-profile ham.

The Challenge of Being a Little Pistol

Looking to snag some big DX stations from your small city lot? Follow the author's tips for DX success.

By Rod Vlach, NNØTT

I must admit, there have been times I wish I had a kilowatt amplifier to bust through a rare DX pileup, but then I think, "Where's the challenge in that?" My station is about as barebones as it gets, and that's the way I like it. I take pride in slowly adding to my DXCC total with patience, a little bit of skill and some luck, rather than 1000 W of amplified power.

For those who may not be aware, a "little pistol" is someone who operates with the "barefoot" power that is built into the radio, generally about 100 W, and a modest antenna. When I bust a pileup, I feel really good about it. I hear the "big guns"—high power stations with directional antennas—blasting away, but more often than not, I eventually get my turn, too, and sometimes I can work the DX before they do.

Here are some hints as to how we "little pistols" can get our fair share of the DX pie, too. Lest someone should think an expensive station and a 100 foot tower with a Yagi antenna is required, I have made all my contacts with a modest—but modern—shack consisting of a medium-priced transceiver, power supply, keying paddles and a vertical antenna. That's it. I make an occasional SSB contact, but I know that with barefoot power, Morse code operating can result in more DX QSOs.

SPLIT MODE AND PILEUPS

Rare DX stations will often operate "split," transmitting on one frequency and listening for responses on another, usually higher, frequency. I have found that I have a good shot at working a DX station that is listening up 2 to 5 kHz from the transmitting frequency, but when it gets crazy, like up 10 to 15 kHz, my chances go way down. This seems to occur mostly

Taking advantage of the odd hour here and there can really pay off—YA5T was worked during a lunch break.

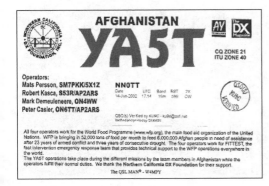

Even the simplest of stations—a small 100 W transceiver going out to a vertical antenna—tucked into a corner of the house can make some amazing DX contacts.

when the DX entity in question is very rare. Not that my signal isn't getting out, but with lower power, I am just one in hundreds of call signs being heard in a 15 kHz range, and it is more a matter of luck than skill if you're noticed by the DX operator. Even when tracking the movements of the DX station, the chances of being chosen are slim—it depends on how much time, patience and effort I think it is worth.

I am particularly proud of my QSL card from VP8THU, a DXpedition station in the South Sandwich Islands in 2002. It's a long way from Minnesota to VP8! The "Micro-Lite Penguins DXpedition Team" announced that they would not be using amplifiers at this rare island, so I knew it would be a tough battle. My contact was on 40 meters using Morse code, and the interference and noise were just terrible. On the back of this card, the QSL manager wrote "Congratulations, Rod." I really appreciated that comment, as that was one of the roughest, toughest rare DX contacts I have ever made, and many operators had to really earn their confirmations. When both the DX and the other station are running barefoot power, a little bit of skill and determination can help!

I do pretty well in head-on pileups, when the DX station is transmitting and receiving on the same frequency. A trick I have learned is that if the pileup is getting intense, use the XIT control (transmitter incremental tuning) to offset your transmit signal up or down about 25 Hz. This makes the signal stand out a little from the crowd, and causes no serious waste of frequency from being exactly on frequency by such a small amount. Besides, not everybody participating in the pileup will be zero-beat with the DX station, anyway. I am surprised how many times this has resulted in a quick QSO for me, while others are hammering away on the main frequency.

BE THE EARLY BIRD

A little sacrifice in terms of time can pay off big when you're DXing with lower power. When the VP6DI group operated from the brand-new DXCC entity of Ducie Island in 2002, the pileups were out of sight. I wasn't getting anywhere in the "up 15" mess, so I decided to get up early and set a trap. I knew by then that VP6DI was sticking mostly to their announced frequencies, so I left my rig on their 15 meter CW frequency the night before. The next morning, I started monitoring. I was one of the first ones to nab VP6DI in that particular run of QSOs. The advantage of being one of the first is that the DX usually listens up only 1 or 2 kHz—or sometimes even stays on frequency—when starting a run, and then increases the listening range

An unexpected catch was 4U1ITU, operated that day by Dave Sumner, K1ZZ.

as the pileup grows. Your chances increase dramatically when the listening range is small.

Another unusual event that happened to me in 2002 was that at the beginning of one of their early morning runs I heard XRØX on San Felix Island send QRL? DE XRØX in Morse. I responded, on frequency, with XRØX DE NNØTT. I was the first to nab a QSO in that particular run, and I imagine the DX operator was surprised to get an instant call without even starting a CQ! Sometimes, acting quickly can help ensure a QSO. It was the first time that I actually had the opportunity to send the call sign of a rare DX station before signing my own call.

Sometimes dumb luck plays an important part in the life of a DXer. All DX-ers sooner or later experience the joy of working new ones they weren't expecting. Although all new DX contacts are fun, I enjoy the unexpected ones the most because of the surprise factor. We can anticipate and hope to work an expedition, but contacting one we didn't know was there is exciting. One that fell into this category for me in 2001 was 4U1ITU, the International Telecommunication Union station in Geneva. It was particularly interesting because ARRL CEO David Sumner, K1ZZ, who was attending an ITU meeting at the time, was operating the station and confirmed the QSO.

Operating out of your normal on-air times can help you find some interesting DX stations you may never have heard before. Due to job requirements and other responsibilities, many of us can usually be on the air only during certain times of day. If you have the flexibility, however, try to occasionally operate out of your normal schedule. You may be surprised how many DX stations are on the air when you usually are not. I have worked some new DX entities during lunch breaks, including YA5T in Afghanistan. Keep in mind that there can be large time differences between your location and the DX station. I have worked many a "midnight QSO"—and sometimes early morning, if I can drag myself out of bed—to nab some new ones. It is unlikely that you will reach the higher DXCC totals unless you have some flexibility in your operating times and take advantage of short opportunities to operate.

THE WEIRD WAYS OF PROPAGATION

I have worked several new DX by checking the upper HF bands, such as 15 and 10 meters, later at night, when they appear to be inactive. By carefully tuning across these bands a few times, I have made several contacts to new DXCC entities, often on the first call. Sometimes, the action is where you wouldn't think it to be. While late night openings on the upper HF bands will decrease as the sunspot cycle wanes, it doesn't take long to spin the tuning dial a few times and it never hurts to check.

Radio frequency propagation via the ionosphere can be the weirdest thing. A band can sound totally dead and then suddenly a signal comes booming in from Australia. Or sometimes, we can hear many hams calling a DX station, but we can't hear the DX operator at all. And sometimes it's the other way around! A band can be dead, and then open at midnight to Europe or the Pacific. The only thing we can be sure of with propagation is that it is not completely predictable. Still, it helps to know a little bit about propagation reports, such as those issued by WWV and found on a number of Web sites, including ARRL's (**www.arrl.org/w1aw/prop**). I'm still try-ing to figure out all the propagation terms, but when solar flux is up, and the K index is down, conditions should result in better HF communications. If I'm looking for a new DXpedition, I often glance at the solar reports to get an idea how "good" the bands are. There were no better conditions for me than when I got active again in 1998, experiencing a few great years of late night contacts on 20, 15, and sometimes 12 and 10 meters, as the sunspot cycle was peaking.

Another indispensable tool for band scouting and evaluation is the NCDXF/ IARU worldwide beacon system on 20, 17, 15, 12 and 10 meters. This coordinated set of propagation beacons will let you know exactly where a band is open to and to what degree. If an upper band is open you'll hear the beacons, including call signs and power levels. Then, if you don't hear anybody having a live QSO, just call CQ a few times—the results can be amazing. See **www.ncdxf.org/beacons.html** for information on how to use the system. Scanning the recent activity on a DX packet cluster can also give you an idea of what bands are open, as well as tell you which stations are on the air.

GOOD EQUIPMENT HELPS

I estimate that my current station cost about $2000—and that includes acces-sories, coaxial cable and everything else. That is modest by some standards, but it is the first brand new station I have ever owned; you can put together a station with used equipment for less. Today's transceivers do a great job and I am pleased with my medium-priced rig. Several very good new 100 W transceivers are avail-able for right around $500, some with digital signal processing. I use a vertical antenna, mainly because I don't want the equivalent of the Eiffel Tower sitting in my backyard. There must be something to the claim that the low angle of radiation from vertical antennas is good for DX. I have heard plenty of great incoming sig-nals originating from inexpensive dipoles and random wires, too, so antenna cost doesn't have to be a big factor. Work within your constraints—but get on the air and work!

It's also important to know your station, especially the capabilities and controls of your radio. Sometimes, you need to act fast if there's to be any hope of nabbing that rare DX station with a lower-powered station. With packet clusters, a pileup can appear very quickly. If you have a modern transceiver, it will be capable of split-mode and incremental transmit and receive. I have learned how to activate these modes swiftly, to have a better chance against the growing "wolf pack." Also know the other features your rig offers, such as noise reduction, IF shift/passband tuning and filters, which can all help you to "hear" that rare one when the going gets tough.

Actually *reading* the operator's manual to discover the full capabilities of your rig is a good idea, too!

HANG IN THERE

As a little pistol, I realize that I won't work every DXpedition or individual DX operator I hear, but my attitude is, "I'll get 'em next time!" There are dry spells, when I won't work a new DX station for several weeks or months (and the higher your DXCC total, the more you will depend on DXpeditions for a new entity). But I'm content to know that my simple station has already contacted the four corners of the globe, and it won't be too long before I nab that next new one.

ARRL member Rod Vlach, NNØTT, and his wife Diane reside in Willmar, Minnesota. He has been licensed since 1965, but got active in 1998. His current DXCC total is 270 on CW. He can be reached at **nn0tt@arrl.net**.

D

Appendix

The following pages are reprinted from the May 2003 issue of *QST*. The article by Ron Herring, W7HD, describes the fixture he made to hold a pair of Hamsticks and use them as a dipole. A dipole made from a pair of Hamstick mobile whips such as Ron's antenna not only is a good portable antenna, but also is a good attic antenna for the low-profile ham.

One suggestion: Because of the difficulty of tuning the two Hamsticks (a problem that becomes greater on the lower-frequency bands), it's a good idea to use an antenna tuner with a double-Hamstick dipole. Another possibility would be to feed a double-Hamstick dipole with ladder line and an antenna tuner, to tune out any mismatch caused by mistuned Hamsticks.

A Small, Portable Dipole for Field Use

The title says it all—a practical, transportable antenna for Field Day and a valuable addition to your emergency kit.

By Ron Herring, W7HD

This antenna came about because I wanted something small and portable that could be used on any band and would perform just like a dipole. Since I didn't want to have to find two trees the right distance apart (difficult to do in arid Arizona) for stringing up a regular dipole, something of a more "stand-alone" nature was needed. Looking through my accumulation of antennas, I discovered that I had a pair of Hamstick mobile antennas for 20 and 40 meters.[1] Needing something for my brand-new PSK20 rig for Field Day, I decided to try building something that would both fit the bed of my pickup truck and be quickly assembled and tuned at a site.

I hit on the idea of using two of these antennas to build a portable dipole. Since a single Hamstick antenna was designed as a mobile antenna and uses the vehicle as a counterpoise, a pair of these appeared ideal for my purpose. I made a quick trip to the local hardware store to pick up some nuts and bolts; a piece of ¾″ by 4″ hardwood (I used oak); a good quality wooden broomstick and some angle-iron with pre-drilled ⅜″ holes. I then proceeded to drill, screw, tape, assemble and make it work. The total cost of materials, including $25 for each of the two Hamsticks, was about $90. Although the Hamstick was available, any suitable shortened (helically loaded) vehicular antenna can be used.

Using nothing more exotic than simple hand tools, a tape measure, power drill, wrenches and screwdrivers, the whole thing came together in about 3 hours. The best part was that it worked exactly as I had planned.

Some tips when you do your own assembly:

- Be sure to tune both antennas on the vehicle before mounting to the assembly.
- For safety, the radiating elements should be out of reach.

---

[1]Lakeview Company, 3620-9A Whitehall Rd, Anderson, SC 29626; 864-226-6990; **www.hamstick.com/**.

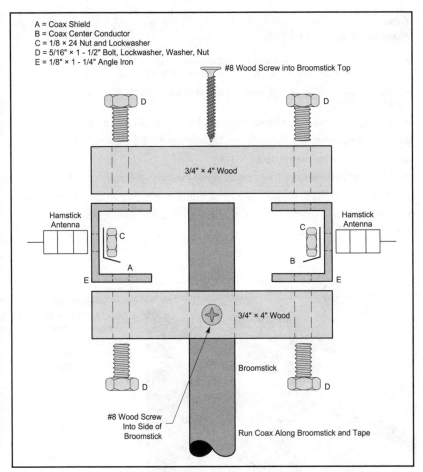

Figure 1—Assembly details for the portable dipole.

• Put a piece of tape on the Hamsticks, marked with the exact length of the "stingers" (the tuning rods) for ease of assembly at the site.

• Treat the wood support with water-seal, lacquer or marine varnish prior to assembly, to prevent deterioration. Just make sure that whatever you use for a coating is non-conductive at RF frequencies.

This antenna will even work on a balcony or supported by a couple of tree branches. My plan is to simply use bungee cords to attach it to the side of the camper at a Field Day site. Since it's a directional antenna, that mounting technique makes it easy to turn. For testing, I simply used bungee cords to attach the antenna to the side of my pickup truck. Table 1 lists the parts necessary to build your own version.

Figure 1 shows how the parts fit together. Figure 2 shows the completed mount. The assembly sequence I used was as follows:

1. Bolt two pieces of angle-iron together to form a "U," making sure that the hole for the antenna is properly aligned. Repeat for the other half.

2. Using the angle iron as a guide, drill two holes in each piece of ¾″ × 4″ wood

Table 1
Portable Dipole Parts list

8—⁵⁄₁₆ × 1½″ bolts with lock washers, flat washers and nuts (nylon-type insert hold best) for mounting the angle iron "U" to the wood pieces.

4—⁵⁄₁₆ × 1″ bolts with lock washers, flat washers and nuts for assembling the angle iron "U" pieces.

4—⅛″ angle iron cut to 3″ lengths (cut so holes line up when mating).

2—⅜″ × 24 nuts with lock washers and flat washers for the antenna mounts.

2—¾″ × 4″ piece hardwood about 5″ long.

2—⅛″ × 1½″ wood screws. (I used decking screws).

2— Mobile antennas that use ³⁄₈″× 24 standard thread mounts (I used Hamsticks).

1—Coaxial cable (I used RG-58/U) stripped and tinned to allow connections about 5″ apart.

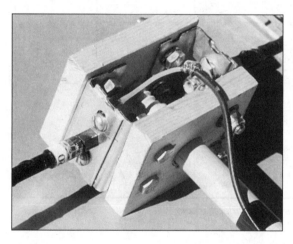

Figure 2—The completed dipole center support showing the broomstick mast, the antenna mounts and the connected transmission line.

support to allow the bolts to pass through. Repeat this on both pieces of wood for each side. Make sure that the gap between the angle-iron pieces is more than an inch, since the broomstick has to pass through the gap.

3. Drill a 1″ hole in the bottom piece of wood for the broomstick to pass through. Additionally, drill a ⅛″ hole in the top piece of wood for a wood screw to secure the top of the broomstick. Drill a ⅛″ hole in the side of the top piece of wood for a second wood screw to anchor the broomstick so it doesn't turn in the mount.

4. Assemble as shown in the figure. Mount the angle-iron "U" pieces to the *inside* of each piece of wood. Be sure to attach the coaxial cable to the metal pieces— I just anchored the wire underneath the lock washers.

5. Stick the broomstick through the bottom hole and put the wood screws in place. (Drill a ¹⁄₁₆″ pilot hole in the broomstick before anchoring, so it won't splinter.)

6. Tape the coax to the side of the broomstick every 18″, leaving the coax free for

approximately the bottom foot of the broomstick.

Okay… it's time to test! Place the antenna in the clear and attach your antenna analyzer or transceiver and SWR meter. Using a low power setting, check the bandwidth of the antenna. It should be about the same as when it was mounted on the mobile mount, perhaps slightly greater. Trim both sides for minimum SWR. Then check the SWR again using full power. Watch for arcing! If arcing does occur, your spacing is too close.

Put a label on the Hamstick giving the length of the "stinger" for the desired operating frequency. You may wish to do this for several favorite operating frequencies. This will save a lot of set-up time at your destination.

Ron Herring, W7HD, has been licensed since 1967 and worked in engineering for the Heath Company in 1968-69 (SB-103, SB-303, MWW-18). While there, he took a two-year course in computer design, which inspired his future. Working for the Kellogg Co (later Michigan Bell) and then Pacific Northwest Bell (Oregon) he also played a role in the development of the RadioShack Model 100 laptop computer. Ron has taught computer classes at Portland State University, as well as a private school. Currently living in Arizona, he works for the Pima County Sheriff's Department in Tucson, where he is a Network Manager. Ron can be reached at 10270 W Mars Rd, Tucson, AZ 85743; **w7hd@arrl.net**.

E

Appendix

The following pages are reprinted from the July 2002 issue of *QST*. In this article, Lew Gordon, K4VX, tells how he shrank a 40-meter dipole for attic mounting by about one-third, making the loaded dipole a good candidate for attic mounting. Lew's loaded dipole is only 45 feet long, as compared to the usual 66 feet for a full-size 40-meter dipole.

The loading sections are built from 450 Ω ladder line, and are supported by lacing the short dipole through the windows of the ladder line. Lew is well known as a successful DXer and Big Fun contester, as well as for his former service to Amateur Radio as the ARRL Director of the Midwest Division.

His home station at the time the following *QST* article was written consisted of 10 towers, ranging from 50 to 170 feet in height, holding up rotary beams for 40, 20, 15 and 10 meters. It's refreshing to see that Lew also thinks of simpler antenna ideas to share with us!

The K4VX Linear-Loaded Dipole for 7 MHz

Here's a very simple and practical way to make a short but efficient 40-meter dipole.

By Lew Gordon, K4VX

One of the easiest antennas to build is the half-wavelength dipole fed with coax. Every *ARRL Handbook* since 1930 contains the information required for constructing one. The only limiting aspect to the antenna is the space, or span, required to support it. This becomes a particular concern at the frequencies 7 MHz and below. While in my own personal situation space is not a problem, I decided to design a shortened dipole for use where space *is* limited.

At 7 MHz a standard horizontal dipole requires approximately 66 feet of wire, and with center and end insulators, approaches 68 feet between rope supports. Obviously one must add several more feet to the total span for support. My concept was to design a dipole that is approximately 70% of the length of a standard dipole, yet be very inexpensive and simple to construct. It also had to perform nearly as efficiently as a full-sized antenna. Linear loading seemed to be the simple way to go.

LINEAR LOADING

Linear loading has been around for many years in the design of 80 and 40-meter

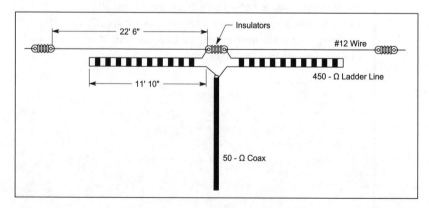

Figure 1—Layout of K4VX linear-loaded dipole. Although the #12 wire is threaded through the 450-Ω window line to support it, this is not shown in this drawing for clarity.

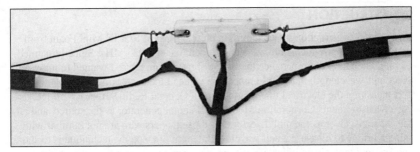

Figure 2—Details of feed point. Note that the feed coax loops through the center insulator for mechanical strain relief.

Yagis. One of the first antennas I can recall was the Hy-Gain 402BA, a 40-meter, 2-element design with 46-foot elements on a 16-foot boom. Other manufacturers copied the concept in their Yagi designs. The shortened dipole presented here is the result of using computer modeling initially to verify the concept, and actual construction to verify the modeling. Not surprisingly theory and reality are very close.

In order to keep the design of this antenna inexpensive I decided that I would try 450-Ω ladder line as the linear-loading mechanism. I pursued two methods of applying the loading:

1. Inserting the loading midway on each side of the span, which requires two additional insulators.

2. Inserting the loading at the center, supported by the antenna wires.

After modeling with *NEC-2*, and later with *NEC-4*, I could see very little difference between these two methods. Since the first method requires two additional insulators, I chose the second method. Both modeling programs provided encouraging results and I decided to construct the antenna.

My initial construction consisted of a 46-foot span of #12 Copperweld fed in the center, and linear loaded with two 12-foot sections of 450-Ω ladder line[1] on each side of the center insulator, and shorted at the end away from the center. Where this design differs from previous ones is that in mine the #12 wire is interlaced into the ladder line to provide physical support to the ladder line. See Figure 1, a schematic of the wiring arrangement. Figure 2 is a photo showing details of the feed point, and Figure 3 shows the final method I used for interlacing the #12 wire through the open windows of the ladder line. I'll discuss this in more detail later.

I can hear the purists now, "That contraption will never work. The wires are too close." Well, this is the proverbial bumblebee—it doesn't know it can't fly!

Figure 3—Final assembly method using interlaced #12 Copperweld wire through holes punched in the windowed ladder line.

CONSTRUCTION

In actual construction I used approximately 54 feet of RG-8X Mini Foam from RadioShack[2] to provide a half-wavelength feed line at 7.025 MHz. Since I assumed I would encounter an impedance that was different from 50 Ω, I wanted to make an exact half wavelength that would repeat the value at the center insulator.

I measured the electrical line length using a General Radio 916-A RF bridge, using a Measurements Corporation Model 65-B signal generator as the source, and an IC-740 transceiver as the null detector. For those readers who are not familiar with using a bridge to measure a half wavelength length of coax, the technique is to short one end and look for an impedance at the other end that is zero reactance and a very low resistive value. In this case after some pruning the resultant impedance was 3.1 + $j0$ Ω.

After the antenna was up in the air, I could then measure the impedance at the feed point and proceed with matching it. As a point of interest, RadioShack states that this coax has a velocity factor of 78%. My measurement was remarkably close, at 77%.

INSTALLATION

Once the antenna was constructed, I hoisted it up to about 40 feet. I then used the MFJ-247 bridge to measure SWR. To my shock, the SWR was perfectly flat at 7.025 MHz, the frequency for which I had designed the antenna! Having been a ham for almost 55 years, I can honestly say that the chances of this happening are almost nonexistent. Figure 4 shows the measured SWR for this initial antenna.

Obviously the design center at 7.025 MHz is too low to cover the entire 40-meter band. However, perusal of the data indicates that the 2:1 SWR bandwidth approaches 300 kHz, which is broad enough to cover the entire band with some scaling of the center frequency. In my own case I would probably have left the antenna as is, but since this is an article for the general amateur population, I proceeded to scale the design to 7.125 MHz.

FREQUENCY SCALING

The first thing learned is that a 1.4% (7.125/7.025) proportional change of both wire length and 450-Ω ladder line length will not provide the correct frequency shift. Originally, I shortened each wire tip 4 inches and each linear loading line 2 inches. This moved the resonance to 7.200 MHz, rendering it less than useful at the bottom of the CW band.

Next, I decided to leave the linear loading alone and just lengthen the ends of the #12 wire. After several attempts, resonance is now 7.125 MHz. Figure 5 shows the SWR of the completed antenna. The 2:1 SWR bandwidth is approximately 275 kHz, which covers practically the entire 40-meter band. The measured final dimensions of the antenna are 22.5 feet of #12 Copperweld, and 11 feet 10 inches of ladder line each side of center.

While conducting these tests, I discovered that raising and lowering the antenna alone could shift resonance as much as 20 kHz. This was with no length changes being made. I then decided to secure the #12 Copperweld as closely as possible to the center of the ladder line with electrical tape. It appears that as the secured Copperweld shifted, the coupling between the ladder line and the Copperweld changed— not much, but enough to be detected by the MFJ-247. Electrical tape alone is not the

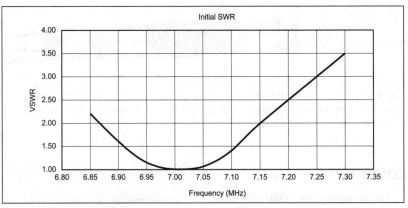
Figure 4—SWR curve for initial model built by K4VX.

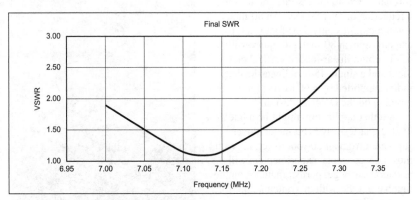
Figure 5—SWR curve for finalized model.

best final solution for this. A better solution is to punch holes in the center of the ladder line insulation and lace the wire through at about 6-inch intervals. This prevents any deviation in the spacing of the conductors. See Figure 3 again.

There are several ways to secure the end of the ladder line to the #12 wire. I chose electrical split-bolt connectors with nylon cord for mechanical strain relief. Care should be taken to insure that the shorted end of the ladder line does not come in contact with the wire. See Figure 6.

PERFORMANCE

There appears to be little measurable on-the-air difference between this shortened dipole and a full-sized dipole. Computer modeling indicates less than 1 dB difference also. Figure 7 is a photo of the antenna at 40 feet.

The total cost of this antenna (minus coax) should be less than $25 with all-new materials. The 450-Ω ladder line is available for approximately $0.25/foot and Copperweld #12 wire for less. New insulators should be less than $10, but junk boxes and flea markets can usually suffice.

Figure 6— There are several ways to secure the end of the ladder line to the #12 wire. I chose electrical split-bolt connectors with nylon cord for mechanical strain relief. Care should be taken to insure that the shorted end of the ladder line does not come in contact with the wire.

CONCLUSION

This design provides a dipole that is 70% of the span of a full-size dipole with little observable difference in performance. On 40 meters this amounts to a reduction of over 20 feet. An 80-meter version requires only a 90-foot span. I can envision an 80-meter 4-square with a 45-foot high linear-loaded vertical element and a single 45-foot linear-loaded radial supported by a 60 to 70-foot tower based upon the ON4UN design.[3]

Another application might be a linear-loaded quarter-wave 160-meter sloper supported from an 80-foot tower. As with most new concepts, experimenters will not necessarily get perfect matches on the first try as I did with this antenna, but for those willing to make adjustments and use some cut-and-try, the rewards will prove worth the effort.

Figure 7— The K4VX 40-meter linear-loaded dipole deployed at 40 feet.

Notes

[1]Available from The WIREMAN, 261 Pittman Rd, Landrum, SC 29356, orders 800-727-WIRE; **cahaba.net/~thewirem/index.shtml**. Part #CQ-552.

[2]RadioShack #278-1313.

[3]*ON4UN's Low-Band DXing* (Newington: ARRL, 1999) p 11-72, Section 5.3.

Photos by the author.

Lew Gordon has been licensed continuously since 1947, and earned his Extra Class license in 1952. He has previously held the calls W9APY, WA4RPK and W4ZCY. He has held K4VX since 1973. Lew's wife holds NSØZ, and his daughter is NØHVY. Lew earned a BS degree in Physics from Purdue University and did graduate work at Georgetown University. Lew is a retired US Government systems engineer. He was elected ARRL Midwest Division Director in 1993 and retired in 2000. An active contester, Lew's main love in Amateur Radio is antenna design and construction. His antenna farm consists of 10 towers ranging from 50 to 170 feet with rotaries on 40, 20, 15 and 10 meters. He has written several human interest articles as well as articles on band-pass filters, antenna designs and other subjects. His freeware program YAGIMAX is in use all over the world. You can contact the author at PO Box 105, Hannibal, MO 63401; **k4vx@arrl.net.**

F

Appendix

The following pages are reprinted from the June 1967 issue of *QST*. In this article, Lew McCoy, W1ICP, describes a small antenna for low-profile use.

This kind of antenna is particularly useful for the apartment dweller, for whom low-profile antennas are a problem. He might live in a building with steel and concrete construction, where indoor antennas won't work well. He may not be able to put a wire antenna out the window. An antenna such as the one described in this article may be an antenna of last resort.

Although this type of antenna is inherently inefficient, it's better than not being able to get on the HF bands at all. And you will definitely be able to make contacts with it (see the discussion of the results N5JED had with his windowsill antenna in Chapter 4, under the heading "My Sixth Choice").

This type of antenna is commercially available nowadays from several manufacturers. The MFJ-1622 is an example of a commercial windowsill antenna.

A Windowsill Semivertical

And Other Antenna Ideas For Apartment Dwellers

BY LEWIS G. McCOY,* WIICP

It is apparent from our mail that a large number of newcomers — and for that matter, old timers — live in apartments or housing developments that have antenna restrictions. A typical question is "What type of indoor antenna can one use in an apartment building that is constructed of concrete and steel?" Or, lacking permission to install an outdoor antenna on the roof, what can one do? This article will treat the problem and present some answers to a rather sticky situation.

Recently, an article appeared in *QST* describing an antenna system for a "traveling man" — an antenna that could be mounted on a windowsill — was easily erected or dismantled, and took up a minimum of traveling space. The only difficulty is that some of the parts used in the system are hard to come by. Fig. 1 and the photographs show a similar system which can be made from easily obtainable parts.

INDOOR ANTENNAS

Let's make one point clear at the outset: Indoor antennas in a steel and concrete building are very inefficient. It is possible to get r.f. power into the antenna, but usually the building tends to block any useful radiation. On the other hand, many amateurs have experienced good, if not excellent, results using an indoor antenna in a frame building. Even so, it should be made clear that in most cases a relatively poor outdoor antenna will outperform an indoor antenna. The problem boils down to how to get an outdoor antenna when the landlord says you can't put one on the roof. If you have access to a window, and most people do, there are several possibilities for getting a "wire" outside.

THE WINDOWSILL SEMIVERTICAL

One such possibility is the antenna shown in the photographs. This system consists of a transmatch and a radiator, the radiating element being approximately 12 feet of Reynolds aluminum tubing. It can be mounted on a windowsill and easily broken down for storage when not in use. We call the aluminum tubing a semivertical because it is mounted at approximately a 45-degree angle from the window.

The transmatch components and the bracket that supports the aluminum tubing are mounted on a 12 × 18-inch piece of plywood. The unit shown was mounted on ¼-inch stock and while it worked out OK, it is recommended that ½-inch plywood be used because it would be a stronger support.

A piece of aluminum, 3 × 15 inches, is required for the radiator support bracket. This is bent up as shown in Fig. 2 and mounted on one end of the plywood piece. Two 1½-inch U bolts are mounted on the bracket to hold the end of the aluminum tubing when the antenna is in use. Enough space should be left on the plywood between the bracket and the coil/capacitor combination to permit the window to be closed. We allowed 1½ inches as this was slightly more than the width of the window sash, which was wood in our case. If your window sash is metal, the lead from the bracket to the end of L1 should be insulated to prevent shorting to the window sash.

You may have to use your ingenuity in coming up with a method for supporting the plywood on the windowsill. In our case, the windowsill protruded into the room about two inches, so an 18-inch length of 1 × 2 wood and two nuts and bolts were used to clamp the system to the sill. This held the assembly securely enough, but the ¼-inch thick plywood was slightly "bouncy." That's why thicker plywood is recommended.

All interconnections on the transmatch were made with clip leads because, as you'll see under tune-up procedures, there are many possible combinations for setting up the circuit and the use of clips makes the job simpler. Both the coil, L1, and the capacitor, C1 are mounted on 1-inch stand-off insulators. A coax fitting, for connecting to the transmitter, is mounted on a small aluminum bracket at one corner of the plywood.

The vertical radiator element is made from one 8-foot length of type 8A and one 6-foot length of type 185 Reynolds Do-It-Yourself aluminum tubing. These two lengths are cut in half to make a storage package with a maximum length of 4 feet. The antenna is made as shown in Fig. 2. Hose clamps are used to clamp the sections together. The ends of the tubing where the clamps are used should be cut back about 3 inches with a hacksaw to permit the clamp to compress the tubing when tightened. These 1-inch clamps are available at any hardware store.

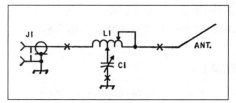

Fig. 1-Circuit diagram of the transmatch for
random length wires.
C1—100-pf. variable; see text (E. F. Johnson
154-14 or similiar).
L1—36 turns No. 14, 2×2 inches diam.,
8 turns per inch (Polycoils 1775, B.&W.
3906-1, Air-Dux 20081).
J1—Coax chassis receptacle (SO-239).

With this type of antenna and transmatch, it is possible to develop some rather high r.f. voltages, even at 75 watts input. While these voltages are not dangerous they can cause a nasty r.f. burn, so be sure to use an insulated knob (no metal skirts) on the rotor of C1. Don't touch any of the metal parts on the transmatch when the system is working.

Because of the high r.f. voltages, receiver type-plate spacing for C1 is not recommended. To be on the safe side, minimum spacing should be on the order of 0.050 inch. The capacitor we used had spacing of 0.125 inch, and at power levels up to 300 watts no arcing was experienced. While a maximum capacitance of 100 pf. is specified in Fig. 1, any value of from 75 pf. to 200 pf. could be used. Many capacitors that fit this category can be found in surplus at bargain prices.

TUNE- UP PROCEDURE

Up to this point, we've only discussed the aluminum tubing as a radiator. However, the transmatch can be used with random-length wires for the antenna. The author's daughter lives in an apartment building on Long Island so we had a chance to test various setups using the transmatch. One antenna was about 60 feet of No. 28 enameled wire suspended from the 20th floor. (For all practical purposes, No. 28 wire is invisible except when one is within a few feet of it.) A small weight was put on one end of the wire and the wire dropped down the side of the building. The other end was attached to the end of L1 and the transmatch adjusted for a match. The side of the building we were

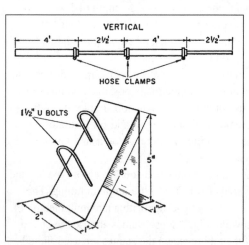

Fig. 2-Top, the vertical radiating element.
Bottom, details for making the mounting bracket
for the vertical.

on faced north, providing a clear shot north with some coverage to the east and west. With about 75 watts input, on 80, 40 and 20 meters, contacts were made in some 30 states over a two-day period. The wire antenna brought better reports than the semivertical from all areas, but the vertical did give a good account of itself. Poorest performance was to the south, where the building was in the way. A 40-foot wire was strung up inside the apartment, but only a few local stations were contacted, and those just barely.

In setting up the system, a reflectometer such as Monimatch2 or Varimatch3 is needed between the transmatch and transmitter to show when the former is correctly adjusted. It is impossible to tell you the exact settings of C1 or where the tap on L1 should be placed, simply because each installation will be different. This difference usually will be in the ground connection. For example, you may be on the 20th story of an apartment building, and for a ground connection you run a lead to a water pipe. There is no way of knowing where this water pipe reaches actual ground, but nevertheless the water pipe is part of the antenna system. On the other hand, there may be no water pipe or anything that resembles a "normal" ground connection at hand. However, there is a ground return — through the a.c. lines. If you can make a good ground connection, fine and dandy. However, don't worry about it; the antenna system will still work without one. The only thing is, we cannot predict what arrangement will be necessary with the transmatch for proper tune-up. You'll have to experiment.

While the adjustment for a given band may be time-consuming for the initial setup, the procedure is quite simple. We suggest you start out with a hookup as shown in Fig. 3, at C. Connect the reflectometer in the coax line between the rig and the transmatch and feed enough power through the system to get a full-scale meter reading in the forward position. Next, switch the

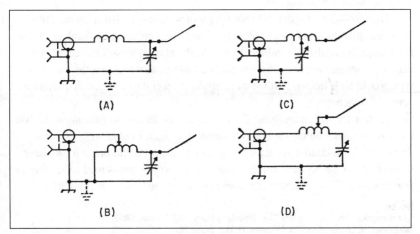

Fig. 3-Four possible configurations for setting up the transmatch to achieve a match.

reflectometer to read reflected power. Tap the stator of C1 on either end of L1. The rotor end should be connected to the metal bracket holding J1. If you have an earth ground, it too should be connected to the bracket holding J1. Tune C1 through its range while looking for a dip in the reflectometer reflected reading. If you don't get a dip, move the stator clip down the coil a couple of turns and try again. At some point along the coil, you should be able to get the reflected reading down to zero. When you have this condition — and on some bands it may mean moving the tap across a fraction of one of the turns — the system is correctly adjusted.

Make a note of the settings so you can return to them, move to the next band, and repeat the adjustments. A chart can be made up with the correct settings for each band, and changing bands then becomes simple and fast. In some instances you may have to try the configurations shown at A, B, or D in Fig. 3. However, you should be able to achieve a complete match with almost any antenna (random wire length) you can come up with.

OTHER USEFUL INFORMATION

Possibilities for antennas where they are not permitted are really up to your own ingenuity. One ham we know had only one window, and that opened on an air shaft. At first glance it would have appeared impossible to get any useful type of antenna outside. However, there was a metal downspot near the window, so he attached a wire to it, used a transmatch similar to the one described here, and ended up with a pretty fair antenna system.

Another attached a wire to the metal screen on his window and used that as an antenna — maybe not the best in the world, but it worked.

Keep in mind that almost any piece of metal, even if grounded, will radiate r.f. With the transmatch described here you should be able to use that piece of metal for an antenna.

One problem with that 60-foot length of wire suspended from the 20th floor was that the wind would blow it around. This changed the loading on the final amplifier of the transmitter. However, the change wasn't radical enough to keep us from making contacts. Probably a little more weight on the end of the wire would have helped, but we were afraid the weight might crack a window several floors below!

As to wire size, hams have used wires as fine as No. 36 for runs up to 100 feet, using rubber bands for end insulators. The main trouble with wire that fine is that birds cannot see it and they fly into the wire and break it. We have found that No. 28 Nylclad copper wire is fairly strong and is hard to see from a few feet away. After all, you can always string the antenna out at night!

Notes

[1]Santangelo, "An Antenna For The Traveling Man," *QST*, April, 1967.
[2]Monimatch, Understanding Amateur Radio, page 209.
[3]De Maw, The Varimatch," *QST*, May, 1966.

G

Appendix

In this book, I've told you about the Killer Antenna that I use as a home-station antenna. Although it was not intended to be a low-profile antenna, it disappears into the woods behind my house quite well. The only thing that's easy to spot is the ladder line that feeds the antenna. First-time visitors to our home will see the ladder line and ask, "What's that for?", never noticing the antenna itself. With a little care in the routing of the ladder line, the entire antenna system would become a low-profile antenna.

The following pages are a reprint of the original article (June 1999 *QST*) that described the antenna.

A Simple Killer Antenna

Forty meters (and more), fifteen bucks—lots of DX. An elevated ground system is the secret to the author's success. Try this antenna at home or during Field Day!

By Al Brogdon, W1AB

Here's an antenna that's inexpensive, easy to build, relatively "invisible," quite compact and a very good performer on 7 MHz (and the higher HF bands). A tree limb about 45 feet above the ground (or a tower) is required for a support, and a good antenna tuner is needed to provide a match between the antenna's 450-Ω ladder-line feed and the rig.

Feeding the antenna with ladder line is the only departure from conventional ground-plane antenna construction. It may at first sound odd, but my expectation was that the 450-Ω feed (1) would result in a low SWR across the entire 40-meter band and (2) would enable me to use the antenna on all the HF bands above 7 MHz. I was not disappointed!

As an aside, I'm a firm believer in using a good antenna tuner and antennas fed with single-wire, open-wire or ladder-line feeders. These antennas have served me well over many years.

My wife Maggie and I moved to the small village of Fitzwilliam, in the southwestern corner of New Hampshire, in July 1998. During the summer and fall I was so busy with all my moving and unpacking chores that I didn't have time to put up any serious antennas. When the air started to get a bit nippy, I began erecting some simple wire antennas to use during the winter. (Permanent antennas—and the tower—would have to wait until spring.)

Quite a few tall trees surround our home, and I planned to use them to support three wire antennas: (1) a 130-foot dipole, (2) a 250-foot end-fed Zepp. and (3) an antenna of some sort that would perform well on 40 meters and the higher bands. I planned to use 450-Ω ladder line to feed each antenna. But first I had to decide what antenna to build for "7 MHz and up."

Jack Colson, W3TMZ, a long-time friend, has a simple but effective antenna for 160 meters—an elevated ground-plane in an inverted L configuration. A 33-foot vertical section rises from the feed point, with a 97-foot horizontal section connected to the top of the vertical piece, for a total length of 130 feet—a quarter wavelength on 160 meters. Jack then added two 130-foot radials, not connected to ground, at the antenna end and about eight feet above the ground. Very simple—and very effective! Jack works just about anyone he hears on 160 meters.

CONSTRUCTION

His success led me to try an appropriately scaled elevated ground-plane with a ladder-line feed as my 40-meter antenna. I cut four 33-foot pieces of wire, using one as the vertical radiating element and the other three as elevated radials.

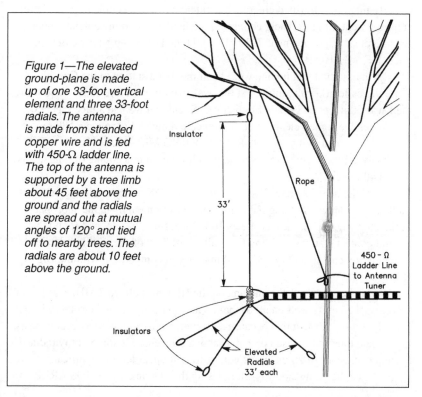

Figure 1—The elevated ground-plane is made up of one 33-foot vertical element and three 33-foot radials. The antenna is made from stranded copper wire and is fed with 450-Ω ladder line. The top of the antenna is supported by a tree limb about 45 feet above the ground and the radials are spread out at mutual angles of 120° and tied off to nearby trees. The radials are about 10 feet above the ground.

Insulator

Rope

33'

450 - Ω
Ladder Line
to Antenna
Tuner

Insulators

Elevated
Radials
33' each

I used a slingshot and a 3-ounce spherical fishing weight to shoot a line over a tree limb to support the antenna. First I shot nylon twine into the tree. When I got the twine over an acceptable limb (about 45 feet above the ground), I used it to pull up a polypropylene rope. (There are better materials to use for antenna-support ropes, but I can get polypropylene rope at my local hardware store.)

I then pulled the antenna's top insulator up to the limb and connected a piece of nylon twine from the base insulator to the ground and tied it off to a "dead man"—a 4-foot length of log— with the twine attached near one end of the log. If the tree holding up the antenna sways and pulls the antenna upward, the antenna will lift the end of the log off the ground, which keeps the strain from breaking the antenna or the top support rope.

I spread the three 33-foot radials out in a roughly equiangular configuration to form an elevated counterpoise—which is not connected to ground—and tied them off to nearby trees using twine. The feed point is about 10 feet above the ground. Because the tree leans toward the antenna, the "vertical" element is, serendipitously, nearly vertical (see Figure 1).

ON THE AIR

By the time I finished erecting the antenna it was a couple of hours before sunset. I went into the shack, connected the ladder line to my antenna tuner, found a clear frequency near 7030 kHz and tuned up. I signed my call sign once: DE W1AB AR.

I had assumed that no one would call me, but a station in the Czech Republic called (keep in mind that this was two hours before sunset) and told me I was 30 dB over S9! His English was fluent and we had a nice ragchew.

When we signed clear, a Greek ham called me and gave me a 589 report, saying (as I noted in my log), "UR ANT DOING MIRACLES!"

Holy smokes! This $15 antenna, which I put up in a couple of hours, was hot on 40 meters.

Over the next few days I continued to work DX easily on 40 meters, including Japan via the long path in the early evening and via the short path in the morning, sometimes as late as 1030 local time. In pileups I found that I could contact stations with my first call about 75% of the time by merely throwing my call sign out once—and that was in competition with the Big Boys.

I soon began trying the antenna on the HF bands above 7 MHz and found that it worked quite well on all of them—harmonically related or otherwise. Frankly, I didn't expect this because a 40-meter vertical tends to develop nasty high-angle lobes when used on bands above 20 meters. If some of my radiation at frequencies above 20 meters is going straight up, at least a significant portion must be going sideways, too. Using the antenna in the 1998 ARRL

DATE	MC	GMT Start	GMT End	STATION WORKED	RST HIS	RST MY	REMARKS	
27 DEC 1998	28	1453	1501	DH2DAM	599	599	RALF	COLOGNE, GERMANY
	7	2051	2052	TL8A	599	599		CENTRAL AFRICA
	14	2057	2122	VK6DX	579	599	DAVE	PERTH, AUS.(LONG PATH)
		2133	2138	TF8SM	579	569	SMARI	ICELAND
28 DEC 1998	7	0218	0221	FM/W8MV	579	599	MEL	MARTINIQUE
		0311	0312	SV1SV/46	599	599		GREECE
		0318	0321	UA9UEO	569	599	MIKE	ASIATIC RUSSIA
		0349	0357	ZS1AAX	599	599	AL	CAPETOWN, SO. AFR.
		0530	0534	9H1GZ	589	579	MARIO	MALTA
		0543	0547	SV2AVP	599	599	THEO	GREECE
		0556	0600	TK/DL7HZ	599	599	TAR	CORSICA
		0608	0619	ON6KD	569	599	KARL	GHENT, BELGIUM
		0620	0624	HB9AAQ	579	579	FRED	HAAG, SWITZERLAND
		0624	0630	DF1TJ	579	579	EUGEN	LÖRRACH, GERMANY
		0635	0640	TK5MP	589	589	TOMASO	BONIFACIO, CORSICA
		0649	0656	DJ4ZA	579	569	HARY	HATTINGEN GERMANY
		0656	0705	DL5DTL	579	589	PETER	DRESDEN, GERMANY
		0705	0713	IN3CKK	579	589	BRUNO	TRENTO, ITALY
		0713	0720	I7MMR	579	559	MAURO	GENOA, ITALY
		0720	0727	DF5ZV	579	579	PETRA (YL)	MARBURG, GERMANY
		0727	0731	DL1SKB	589	599	HAR	SCHWEIRN, GERMANY
		0731	0739	DL6SXI	579	599	KARL	PLAU, GERMANY
		0739	0745	DJ7YT	589	589	JÜRGEN	PLOEN, GERMANY
		0745	0752	DK7JJ	459	589	BERT (7W)	KÖLN, GERMANY
		0752	0755	F6DEF	579	579	JACK	FRANCE
		0816	0820	PP7JR	579	569	JOAO	MACEIO, BRAZIL
		0823	0824	PA3EDO/PEU1K	579	599		NETHERLANDS
29 DEC 1998		2112	2115	GW0GEI	589	599	STEVE	WALES
		2119	2122	JA3XNH	579	569	MIO	JAPAN
		2132	2154	DL2QB	579	569	KLAUS	COLOGNE, GERMANY

Figure 2—An extract from W1AB's log for December 27-29, 1998.

10-Meter Contest, I worked DX all over the world, including New Zealand. And I got nothing less than 599 reports (that's a joke, by the way. considering that everyone seems to get 599 reports during contests these days).

Even on the competitive 20-meter band I found that I could work through pileups easily with 100 W, often beating out stations with Serious Antennas and High Power. One evening when 20 meters appeared to be dead for DX, I had a nice chat with longtime friend K4LTA in Tennessee. After we finished, a Japanese ham called me and gave me a 599, telling me that he was beaming via the long path. Then a ham in Norway, north of the Arctic Circle, called in with a 599. Then HS98AG (Thailand's special event station commemorating their hosting of the Asian Games) called and gave me a 599. Please note that these stations called me, rather then my having to chase them down. And even in outrageous 20-meter pileups, the antenna makes me quite competitive!

After sunrise one morning, I tuned up on 80 meters for some ragchewing, but forgot to switch from the 40-meter ground plane to my 80-meter dipole. I got a 599 (an actual report, since this wasn't in a contest) from a station in the Deep South. It surprised me that the antenna works so well on 80, but I'll take the unexpected bonus!

Conditions in the 1998 ARRL 160-Meter Contest were only fair, and I had to work hard for contacts using my 250-foot end-fed Zepp. At one point I had called a Wyoming station for several minutes on the Zepp—with no success.

Then I called him using my 130-foot dipole (a fair performer on 160 meters)—still no reply. What the heck—I tried the 40-meter ground plane (at reduced power to avoid arcing in my tuner)—and got him on the first call! The antenna shouldn't work very well on 160 meters, but it made the long-haul contact when my other—supposedly better—antennas didn't. Go figure!

During the few days following Christmas 1998, I spent a bit more time than usual on the air. Figure 2 is an extract from my log for some of the contacts during that period—mostly on 40 meters. Note the time period from 0649 to 0755 UTC on December 28; every time I would sign clear from one contact, between two and a dozen other stations would call! On December 29 at 2119, you see another long-path Japanese contact (Mio has a very big signal via 40-meter long-path propagation on a regular basis).

I can't promise that you will experience similar results—as the car dealers say, "your mileage may vary"—but you might be surprised at how well you can get out with this simple antenna.

Several hams have asked me about the vertical element being so close to the vertical tree trunk, and what losses that proximity might induce. Even when the trees were most conductive (when the sap was running), the antenna worked quite well. If there was any loss, it wasn't enough to bother my DXing!

It's a pleasure to see how well a vertical antenna works when using a 450-Ω ladder-line feed. I decided to leave my killer vertical in place when going to my "permanent" antenna field!

The author's elevated ground-plane antenna is shown on the right side of the frame. You can see the ladder line, but the antenna doesn't show up very well because it's practically invisible. Even if you can't see the antenna, you can appreciate the view of Mount Monadnock!

H

Appendix

Several years ago, Albert Parker, N4AQ, moved into a retirement community that had deed restrictions against any kind of antenna. He wanted to use a Hustler 4-BTV vertical as his antenna. Inspiration hit him—he placed his 4-BTV inside a piece of PVC tubing to conceal it, and then used the PVC tubing as a flagpole. Let's run that idea up the flagpole and see who salutes...from May 1993 *QST*.

A Disguised Flagpole Antenna

Is it a flagpole or an antenna?
Actually it's both.

By Albert Parker, N4AQ

I moved to a retirement community
in October 1991—a community which has
restrictions against erecting any sort of antenna.
I was eager to get on the air and began looking
around for an inconspicuous antenna. It couldn't
be just any antenna; I needed one with multiband
capability. Many multiband antennas use tube-like
assemblies made of coils and capacitors. These
are known as traps and they electrically separate
one part of the antenna from another, depending
on the frequency of the transmitted signal.
Traps do their jobs well, but they send a clear
message when they're displayed in public: HAM
ANTENNA HERE!

I was visiting a local ham dealer one day
when I noticed a Hustler vertical antenna on the
roof. I took a closer look and began to wonder
how I could hide such a thing. Suddenly, it
occurred to me that the antenna— including the
traps—was thin enough to fit inside a 2-inch diameter PVC pipe. Hmmm, an
enclosed antenna would not only radiate well, it could serve as a flagpole to
disguise its real function!

CONSTRUCTION BEGINS

I purchased the Hustler 4-BTV, a four-band trap vertical antenna. (I'm
sure a five-band vertical would work just as well for my purposes but, of

course, it would be taller.) As soon as I got it home, I took one of the trap assemblies to the local hardware store and tried to squeeze it inside a 2-inch PVC pipe. The trap's hose clamps were too large to fit! Knowing nothing about plumbing and even less about PVC pipes, I was very discouraged.

I journeyed to another hardware store that carried all sorts of plumbing and sprinkler supplies. I found another 2-inch section of PVC and again tried to insert the trap. To my great delight, it fit perfectly! This was thin-wall PVC. The first one I tried was heavy-duty PVC. Maybe there was hope after all!

I decided to go full blast with the project and purchased a 14-foot section of 2-inch thin-wall PVC along with a 12-foot section of 1½-inch PVC. I assembled the Hustler and cut the 2-inch PVC to fit over the lower part of the antenna, ending about two inches above the 20-meter trap. At that point I used a reducer to couple to the 1½-inch PVC I had slipped over the thinner top section that remained. When I finished, the entire antenna was enclosed in PVC! The 40-meter section used an assembly called a capacitance hat, but I had to leave it off. I later found that I had no trouble operating on 40 meters without it.

I drove a 1½-inch thick-wall pipe about 4 feet into the ground to serve as the base of the antenna. I trimmed the length so it would keep the feed point about 4 inches above the ground. I also drove a five-foot section of copper pipe into the soil to serve as my ground connection. This was hardly an ideal ground for a vertical, but it was impossible to bury a bunch of radial wires in the lawn without attracting unwelcome attention!

I fed the antenna with 50-Ω coaxial cable (discreetly buried, of course) and used about 10 turns of coax near the base to act as an RF choke. I placed a similar choke near the transmitter. Using my MFJ 989C antenna tuner, I was

The base of the flagpole antenna. Note the copper pipe ground connection and the flowers strategically placed to hide the coax. (photos by Albert Parker, N4AQ)

pleased to see that the antenna loaded very well on all bands.

Finishing touches consisted of adding a used toilet-tank-float ball on top of the pole as an ornament, and a three-inch bolt near the top to mount a pulley. The base is hidden by a group of flowers, which I water often to enhance my ground conductivity! My neighbors see the Stars and Stripes flying proudly day after day, unaware that the flagpole is really a multiband vertical antenna.

CONCLUSION

My disguised antenna works well on all bands from 40 meters through 10 meters. I use a Collins KWM-380 transceiver and get good reports on both CW and SSB. I can't compete with the big boys and their high-gain antennas, but it's a perfect alternative if you live in an area with tight antenna restrictions...and it makes a beautiful flagpole!

If you'd like more details on the materials or construction, I would be delighted to help. When you write, however, please enclose a self-addressed, stamped envelope.

Born in Bonneau, South Carolina, Albert received his ham license in 1949 as W4AX. He spent 25 years in the Navy, retiring as a Master Chief Communications Technician. After leaving the Navy, Albert spent two years in Turkey before moving to Saudi Arabia for three years while working in the Lockheed C-130 program. When he upgraded to Extra Class, Albert received his current call sign, N4AQ.

I

Appendix

Mark Weaver, WB3BJF, built a multi-wire vertical antenna, placing the various wires (one for each band he wanted to operate on) around the trunk of the tree. The wires virtually disappear in this configuration.

Remember, if you want your "tree" support to sound official, call it by its military nomenclature, pronouncing it as "tee-are-double-e."

A Four-Band "Tree" Vertical

If the thought of a high-visibility HF antenna leaves you cold, it's time to branch out and get to the root of the problem...

By Mark Weaver, WB3BJF

I live in a townhouse on a small lot in a neighborhood where no outdoor antennas are allowed. That's a fairly typical situation these days. So if I want to operate on the HF bands, am I resigned to an attic dipole or some other indoor compromise? No way! Believe it or not, I'm the proud owner of a four-band full-sized vertical antenna, and it's sitting right in my front yard. And the best part of all is the fact that my antenna is virtually invisible. No Klingon/Romulan cloaking devices here, just old-fashioned ingenuity.

THE CONCEPT

I tried an attic dipole and had nothing but problems. RF got into everything! It got into the TV and the kids howled. It got into the telephone and my wife howled. The antenna also picked up every kind of noise from my computer, TNC and any other electronic devices in the house.

One day while staring out my front window, dreaming of 100-foot towers and stacked Yagis, my gaze fixed upon a solitary 20-foot tree in my front yard. Wait a minute! I can run a 15-foot hunk of wire up the side of that tree! That's almost a quarter wavelength on 20 meters! But what about 40 meters, one of my favorite bands? I decided to worry about that later. Thus was born my four-band "tree" vertical.

My idea isn't new, although the application may be unique. The antenna is comprised of three quarter-wavelength wires (for 10, 20 and 40 meters), snaking up the side of the tree, more-or-less in parallel, all soldered together at the bottom to the center conductor of the coax (see Figure 1). Several radials are then soldered to the ground braid of the coax. But how do you get a quarter wavelength wire for 40 meters into a 20-foot tree? That's over 30 feet of wire! Easy. Bend the wire at the halfway point and run it to an upstairs window of your house, or some other convenient support. When you're finished you'll have an inverted L on 40 meters, a vertical with the top bent over so it looks like an upside down L. The 40-meter inverted L also works on 15 meters, where it's ¾ wavelength.

CONSTRUCTION

Cut three pieces of wire at quarter wavelengths on 10, 20 and 40 meters using the formula:

$$\text{Length (feet)} = \frac{234}{\text{Frequency(MHz)}}$$

Choose frequencies that are in the middle of the bands, or your favorite band segments. I recommend #26 enameled wire available at RadioShack. It's strong and difficult to see.

Examine your chosen tree and the surrounding area. Make absolutely certain that you're not near any power lines. If you see power lines running through the branches, find another tree.

Once you've selected your tree, run the wires up the side of the trunk. If you're an experienced tree climber, work your way up the branches taking the wire along as you go. The alternative is to use a ladder, but make sure you have someone on the ground hold it for you. If inquiring minds want to know what you're doing, explain that you're trimming the tree and/or inspecting the leaves. You can attach the wires to the tree with loops of fishing line, or any other lowvisibility means.

The idea is to get the wires as high as possible. You'll probably have to bend the 20 and 40-meter antenna wires, depending on the height of your tree (and your desire to climb it!). The tree in my yard is about 20 feet from the house, so the end of my 40-meter antenna reaches to an upstairs bedroom window. If you use your house as a support, you need to insulate the antenna wire so it won't come in contact with metal siding, storm windows or whatever. I loosened a screw in my metal window frame, tightened it down on a piece of insulated wire and tied the antenna to the wire.

Now build your ground system. Take bare copper wire, preferably something thick like #14, and loop it around the bottom of the tree at ground level. Solder several radial wires to this ioop and run them out into the yard. I only used six radials at various lengths, making each one as long as possible. Bury the radials about an inch beneath the soil. (Do this at night if you live in an antenna-restricted area. If anyone asks, just tell them you're checking the lawn for grubs.) The radial wires don't have to travel in straight lines. Zigzag them as much as necessary to fit the available space.

Now install the transmission line. My townhouse, like many, has a water spigot on the front. That means there is a hole through the house for the water pipe to pass through. There was enough extra room in this opening to pass a length of RG-58 coaxial cable. You may need to bury this coax between the tree and the house, so make sure to buy cable that's made specifically for burial in soil.

Back out at the tree, solder all three antennas to the center conductor of the coax and solder the coax braid to the copper radial loop. Weatherproof the coax connections. I used RadioShack "Outdoor RF Connector Sealant" (part no. 278-1645). Cover the copper radial loop with mulch or soil so it won't be visible. I planted pansies around mine and it looks very nice.

TESTING

An SWR meter is all you need for testing. If you measured the antenna lengths correctly, the SWR will probably be no higher than 2:1. If you want it lower, add a

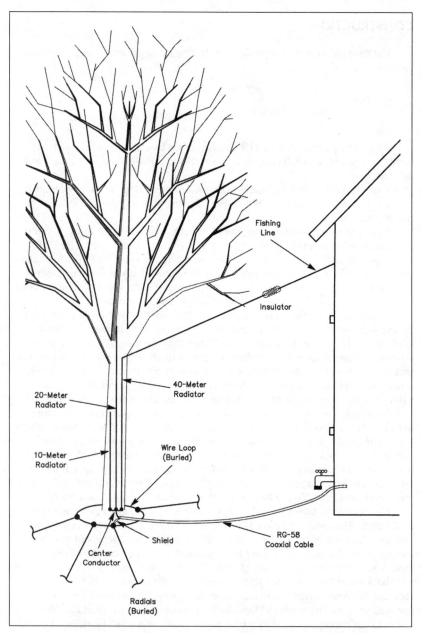

Figure 1—Run the three antenna wires along the trunk and then, if necessary, bend them along the branches. My 40-meter wire is so long that it leaves the tree altogether and attaches to my window frame. Use at least six radials for your ground system, more if you have the space and the patience to place them beneath the soil.

few inches of wire to the antenna, or trim it as the case may be. If you have an antenna tuner, you don't have to bother with tuning unless the SWR is grossly out of whack. Simply adjust the tuner for a flat 1:1 SWR. Because your transmission line is likely to be short, an elevated SWR isn't as bad as it seems. On 15 meters you're using the 40-meter antenna on the third harmonic. This means that your SWR might be high, but the tuner should be able to take care of it.

RESULTS

Is this antenna "optimal?" Far from it. No doubt there is some RF absorption by the tree, and the radiation patterns probably look like abstract art. I'm sure that some RF is being used to heat the coax when the SWR is high.

The point, however, is that this antenna solved my problems. It works well and is far enough from the house that I no longer have complaints about TVI and telephone interference. Signals from my computer and TNC are but distant memories.

Many operators are astonished when I describe my antenna. They can't believe that my signal is so strong. When conditions are decent, I even work a fair amount of DX. I've also managed to use the system on 30 and 17 meters with good success.

As far as visibility is concerned, you can't spot the antenna unless you walk right up to the tree. Even then, you need to know what you're looking for. So far it's been completely disregarded by the spies from the homeowner's association.

Take it from me: If you live in an apartment, townhouse or condo, you can get on the HF bands with a full-sized antenna. If you can see a tree anywhere on your lot, you've just found a home for your next antenna—and it will probably outperform any indoor design. I must admit, however, that I still stare out the window and dream of 100-foot towers and stacked Yagis!

Notes

Notes

Notes

Notes

Notes

FEEDBACK

Please use this form to give us your comments on this book and what you'd like to see in future editions, or e-mail us at **pubsfdbk@arrl.org** (publications feedback). If you use e-mail, please include your name, call, e-mail address and the book title, edition and printing in the body of your message.
Also indicate whether or not you are an ARRL member.

Where did you purchase this book?
  ☐ From ARRL directly   ☐ From an ARRL dealer

Is there a dealer who carries ARRL publications within:
  ☐ 5 miles   ☐ 15 miles   ☐ 30 miles   of your location?   ☐ Not sure.

License class:
  ☐ Novice ☐ Technician ☐ Technician Plus ☐ General ☐ Advanced ☐ Extra

Name _____ ARRL member?   ☐ Yes ☐ No
_____ Call Sign _____
Daytime Phone   (   ) _____ Age _____
Address _____
City, State/Province, ZIP/Postal Code _____
e-mail address _____

If licensed, how long? _____

Other hobbies _____

Occupation _____

For ARRL use only	LPAR
Edition	2 3 4 5 6 7 8 9 10
Printing	1 2 3 4 5 6 7 8 9 10

From _____

_____

_____

EDITOR, LOW PROFILE AMATEUR RADIO
AMERICAN RADIO RELAY LEAGUE
225 MAIN STREET
NEWINGTON CT 06111-1494

——————————————— please fold and tape ———————————————